International Cooperation for Education About Standardization (ICES 2018)

Strengthening Industry and Engineering, Science, and Management Education Through Standardization Learning

Joint International Conference with 5th Annual Conference on Industrial and System Engineering and World Standard Cooperation Academic Day

SHS Web of Conferences Volume 49 (2018)

Yogyakarta, Indonesia
3 – 5 July 2018

Editors:

B. Purwanggono
D. Hatta Fudholi

ISBN: 978-1-5108-7248-6

Printed from e-media with permission by:

Curran Associates, Inc.
57 Morehouse Lane
Red Hook, NY 12571

Some format issues inherent in the e-media version may also appear in this print version.

Printed by Curran Associates, Inc. (2018)

For additional information, please contact EDP Sciences – Web of Conferences at the address below.

EDP Sciences – Web of Conferences
17, Avenue du Hoggar
Parc d'Activité de Courtabœuf
BP 112
F-91944 Les Ulis Cedex A
France

Phone: +33 (0) 1 69 18 75 75
Fax: +33 (0) 1 69 28 84 91

contact-edps@webofconferences.org

Additional copies of this publication are available from:

Curran Associates, Inc.
57 Morehouse Lane
Red Hook, NY 12571 USA
Phone: 845-758-0400
Fax: 845-758-2633
Email: curran@proceedings.com
Web: www.proceedings.com

TABLE OF CONTENTS

Statement of Peer review

In submitting conference proceedings to *Web of Conferences*, the editors of the proceedings certify to the Publisher that

1. They adhere to its **Policy on Publishing Integrity** in order to safeguard good scientific practice in publishing.
2. All articles have been subjected to peer review administered by the proceedings editors.
3. Reviews have been conducted by expert referees, who have been requested to provide unbiased and constructive comments aimed, whenever possible, at improving the work.
4. Proceedings editors have taken all reasonable steps to ensure the quality of the materials they publish and their decision to accept or reject a paper for publication has been based only on the merits of the work and the relevance to the journal.

Title, date and place of the conference: 2018 ICES JOINT INTERNATIONAL CONFERENCE, YOGYAKARTA, INDONESIA

Proceedings editor(s): BAMBANG PURWANGGONO

Date and editor's signature: Aug. 13 / 2018

Green building standard assessment at a higher education institution

Rahmat Nurcahyo[1,*], *Nurmala* Sari[1], *Muhammad* Habiburrahman[1] and *Ellia* Kristiningrum[2]

[1]Department of Industrial Engineering, Faculty of Engineering, Universitas Indonesia Kampus UI Depok 16424, Indonesia
[2]National Standardization Agency for Indonesia, BPPT 1st building, Jakarta 10340, Indonesia

Abstract. The environmental impact caused by the construction of the building is very large. A building will require great energy, electricity, water, and also produces large amounts of waste. To prevent that, a concept of development that caring the state of the environment is needed. The concept is known as the green building concept. This research was conducted to assess academic buildings at a higher education institution using green building assessment. The criteria for assessment is based on Greenship Rating System which consists of six categories such as appropriate site development, energy efficiency and conservation, water conservation, material resources and cycle, indoor air health and comfort, building and environment management. The results of this research are comparison between the green building actual and ideal conditions using gap analysis. Action plans also developed to achieve the expected green conditions according to Green Building Certification.

1 Introduction

The current global warming is a world issue. The phenomenon of global warming is the rising temperature of the earth's surface due to the increased greenhouse effect. Based on data from the World Green Building Council, throughout the world, building donate 33% of CO2 emissions, consume 17% water, 25% of wood products, energy use by 30-40% and 40-50% utilization of raw materials for the construction and operation. [1] Energy consumption in the building sector can be quite large for lighting or air conditioning. A source of energy for the building comes from the consumption of fossil fuels that emit carbon dioxide in significant quantities. In addition, the construction sector also contributed greatly to the consumption of water and electricity in a region.

Buildings also have a significant impact on human health. Indoor air typically contains between 2 and 5—and occasionally greater than 100—times more pollutants than outdoor air. As a result, poor indoor air quality in buildings has been linked to significant health problems such as cancers, asthma, Legionnaires' disease and hypersensitivity pneumonitis [2]. One idea that to reduce global warming and environmental degradation is an environmentally friendly building concept, called green building. Green building has many

* Corresponding email: rahmat.nurcahyo26@gmail.com

terms such as "sustainable development" s (real estate industry), "green building" (US term), "sustainable building" (UK and Australia term), "sustainable architecture" and "sustainable construction". [3]

Green Building Council Indonesia (GBCI) is an independent institution (non-government) and non-profit (non-for-profit) which one of the programs is organizing the Green Building Certification activities in Indonesia by Indonesian special assessment tools called Greenship Rating System. Several previous research related to the implementation of Green Building in Indonesia such as on Civil Engineering Institut Teknologi Surabaya (ITS), building H ITS and construction project in Surabaya. [4]

2 Green Building

The implementation of Green building concept is not only used in design and construction stage but also the operation stage [5].According to The United States Environmental Protection Agency (USEPA), green building is the practice of creating structures and using processes that are environmentally responsible and resource-efficient throughout a building's life-cycle from siting to design, construction, operation, maintenance, renovation and deconstruction. This practice expands and complements the classical building design concerns of economy, utility, durability, and comfort. Green building is also known as a sustainable or high performance building. [6]

According to Green Building Council Indonesia (2010) [7], the green building is a new building that is planned and carried or a building that already built which is operated by taking into environmental factors or the ecosystem and meet the performance such as appropriate site development, water saving, energy saving, waste reduction, indoor air quality. The benefits of green building are related to environment, economics and social. [8]

There are three types of green building benefits such as environment, economic and social. Example of environmental benefits of green building are reduction of harmful gas emissions and waste. Economic benefits of green building such as energy and water savings, increased property values, increased property values. Social benefits of green building such as improved health, and healthy lifestyle [9]. The benefits are achieved through some approaches such as using solar panels, public transportation facilities provision, stormwater and greywater management.

Greenship Rating System is a tool to assess a building based on green building concept. The rating system containing a grain of aspects of the so-called rating and assessment of each item has a point value or credit point. There are several types of the rating system such as the rating system for new building, existing building, interior space, homes, and the draft for sustainable neighborhood. In this research, assessment tools used is for the existing building.

There are six assessment categories on greenship rating system for existing building, such as appropriate site development, energy efficiency and conservation, water conservation, material resources and cycle, indoor air health and comfort, building and environment management. Each assessment category has several prerequisite and criteria which related to scoring. Some categories only have maximum scoring while other categories have maximum and bonus points.

Appropriate site development (ASD) category have two prerequisites and 8 criteria with 16 points maximum. Energy efficiency and conservation (EEC) category have two prerequisites, five main criteria with 36 maximum points and two bonus criteria 8 bonus points. Water conservation (WAC) category have one prerequisite, seven main criteria with 20 maximum points and one bonus criteria with 2 bonus points. Material resources and cycle (MRC) category have three prerequisites, five main criteria with 12 maximum points.

 https://doi.org/10.1051/shsconf/20184901001

Indoor air health and comfort (IAHC) have one prerequisite and eight criteria with 20 maximum points. Building and environment management (BEM) have one prerequisite and five criteria with 13 maximum points.

3 Research Object and Method

One of the important building in higher education institution is academic building usually being used for classroom activities. Academic building is one of the buildings which use substantial energy. This is caused by the demands of academic services and comfort in the learning process, including the beauty of the room (lighting systems), the comfort of air (air system), complete facilities and others. The necessity to reduce energy use by academic building become the background of this research.

This research objective is to apply green building assessment criteria on the academic building at Faculty of Engineering, Universitas Indonesia. The assessment purpose is to determine the level of application for the green building concept on these buildings and as a reference to achieve green building certification.

There are two steps for assessment of the actual state of the building to the standards Greenship Rating System. The first step is observing the actual conditions that exist in the building that are fit to the items in the category of greenship existing building. The second step after reviewing the actual condition of the buildings is awarding of points for each category that is adjusted to the existing standards in the category of greenship existing building. After the assessment, the analysis carried out for each criteria in the category of greenship existing building along with the action plans that are suggested to meet the standard points.

4 Results and Discussion

There are two buildings at Faculty of Engineering Universitas Indonesia that selected as the objects of this research. The two buildings are academic buildings. The result of the assessment of the first building is 38 points out of 117 points in total so that the percentage of green condition is 32%. The result of assessment of the second building is 20 points out of 117 points in total, so that the percentage of green condition is 17%. Detail of assessment result for each criteria for the two buildings is shown in table 1.

Table 1. Results of Assessment Points

Criteria	Maximum Points	The First Building	The Second Building
Appropriate Site Development (ASD)	16	4	4
Energi Efficiency & Conservation (EEC)	36	21	6
Water Conservation (WAC)	20	2	1
Material Resources and Cycle (MRC)	12	4	4
Indoor Health and Comfort (IHC)	20	3	3
Building Environment Management (BEM)	13	4	2
Total	**117**	**38**	**20**

The analysis carried out for each category in Greenship Existing Building along with the action plans that are suggested to meet the maximum points. Based on the data of ASD criteria, the gap points are caused by several factors, including no restrictions on the use of motor vehicles, the number of bicycle parking and facility doesn't meet the GBCI standard, the formation of plants in the green open land is not accordance to Permen PU No. 5/PRT/M/2008, clause 2.3.1, the lack of a system for treating stormwater runoff, the absence of Standard Operation Procedure for pest and weed control, unavailability of non-domesticated habits and animals.

The gap points at EEC criteria are caused by several factors, including re-commissioning have never done on the cooling system, the lack of written Standard Operation Procedure for air conditioning systems, MVAC system is not used, no submetering for energy consumption in EC buildings.

Based on the data of WAC criteria, the gap points are caused by several factors, including the absence of Standard Operation Procedure for the plumbing system on a regular basis, no historical data of water use because the new water metering tool is just installed, unavailability of water recycling system which output is equivalent to the clean water standards.

The gap points at MRC criteria are caused by several factors, including still using R22 in air conditioning systems which ODP value = 0.05, unavailability of reports for used goods delivery.

Based on the data of IHC criteria, the gap points are caused by several factors, including the lack of measurement of the quantity and quality of indoor air levels to meet the requirements of SNI (Indonesia National Standard) and Permenkes (Indonesia Ministry of Health), no survey for the level of convenience to building user as a means of feedback for improvement.

The gap points at BEM criteria are caused by several factors, including innovation applications to improve the quality of the building is still needed to be developed further, unavailability of Standard Operation Procedure and training which include efforts to meet the criteria, unavailability of Greenship Professionals involved in operations and maintenance.

5 Conclusion

This research was conducted to assess academic buildings at Faculty of Engineering Universitas Indonesia using green building assessment. There are two building selected as the objects. The results of achievement points on the first building measurement is 32%. The results of achievement points on the second building measurement is 17%. Minimum points to obtain a certificate Greenship Bronze level is 35%, so there is a gap condition in both buildings.

The action plan needs to be done to achieve green building certification at least 35% points, among others:

- Appropriate Site Development: An effort to reduce the use of private vehicles, making decent support facilities to encourage the users building to use the bike, utilization of vacant land to be planted with vegetation, making systems for treating stormwater runoff.
- Energy Efficiency & Conservation: Installation of metering tools in EC building for controlling electrical energy consumption, establish the Standard Operation Procedure for air-conditioning system covering operation, maintenance, and reporting.

- Water Conservation: Establish the Standard Operation Procedure for piping systems periodically to prevent leakage and wastage of water, study for water recycling system that could be used for irrigation and flushing toilets.
- Material Resources and Cycle: Replace refrigerant materials with non-CFC material, establish Standard Operation Procedure which governs the way the distribution of second-hand goods which still can be recovered.
- Indoor Health and Comfort: Measurement for air quality in the room at regular intervals, as well as measurements of dust levels.
- Building Environment Management: Socialization and training related to the concept of green building so that every employee, staff and building occupants understand clearly the application of the concept of green building so it is easier to make and understand the Standard Operation Procedure include efforts to comply with the criteria in greenship.

Acknowledgement

Supported by Hibah PITTA 2018 Universitas Indonesia, contract number 2513/UN2.R3.1/HKP.05.00/2018.

References

1. M. Samer. CIGR Journal. **15**, 2 (2013).
2. M. Baum. Green Building Research Funding: An Assessment of Current Activity in the United States, http://www.usgbc.org/Docs/Archive/General/Docs2465.pdf. (2006).
3. A. Vinyangkoon. Commercial Green Building The Value Added Gap Perspective. Thesis, Department of Real Estate Management (2012).
4. F. N. Wakhidah., C. Utomo. Jurnal Teknik Pomits. **3**, 2 (2014).
5. T. S. Surjana. Jurnal Arsitektur Universitas Bandar Lampung. **2**, 3 (2013).
6. R. Zigenfus. Element Analysis of the Green Building Process. Thesis, Rochester Institute of Technology, Rochester, NY, (2008).
7. Green Building Council Indonesia. Greenship untuk bangunan baru versi 1.2. Ringkasan kriteria dan tolak ukur. Retrieved from: http://gbcindonesia.org/greenship/rating-tools, (2014).
8. N. S. A. Adiwoso, Prasetyoadi, S. Perdana. Towards Indonesia Sustainable Future through Sustainable Building and Construction (2010).
9. S. D. Sentman. The Journal of Israel Innovation, **1** (2009)

Quaternary quality a medium quality that connects macro quality with micro quality

Mingshun Song[1,*], *Tingting* Zhu[1] , and *Binfang* Qi[1]

[1]China JiLiang University, Hangzhou, Zhejiang, China

Abstract. To construct the outlook on great quality, the concept of quality can be divided into two aspects, micro quality and macro quality at present. However, the definition of micro quality and its contents are not related organically with the conception of macro quality and its contents. So an intermediate variable between micro quality and macro quality is needed to link them together, and this intermediate variable is named as medium quality. This paper introduces the connotation of medium quality that includes the concept of quaternary quality, namely zero quality, unary quality, binary quality and ternary quality, and explains the progressive logical relationships of these four qualities. Micro quality and macro quality are cohered by medium quality organically, and thus formulates the scientific outlook on great quality.

1 Introduction

Nowadays many countries attach great importance to the quality of economic development, which is defined as development quality in this paper. Therefore, one of the most important goals for China to advance along the path of scientific development is the pursuit of development quality. Development quality, which consists of core factors including reputation, market rules, market competitiveness and brand, is the degree for economic development to meet the requirements of the whole society and sustainable development. Development quality, also known as macro quality, is one of the hottest topics in quality academia in recent years[1]. Product quality, which includes hardware product quality, software product quality, processed material quality, service quality and project quality, is the traditional research field in academia. Relative to development quality, product quality is also called micro quality. The concept of quality is divided into micro and macro aspects, and thus formulates the outlook on great quality[2]. According to the research paradigm, micro quality and macro quality shall have a logical relationship, but now this relationship is ambiguous. Quality practitioners and academia still focus on the concept of product quality, and the definition of product quality mainly based on the International Organization for Standardization (ISO), that is, "Degree to which a set of inherent characteristics fulfils requirements." This definition can be interpreted as the degree that product characteristics conform to standards and meet the requirements of customers[3]. The formation of the outlook on great quality must be based on the combination of micro quality and macro

* Corresponding author: smsqm@cjlu.edu.cn

quality, that is, macro quality embeds micro quality, and micro quality determines macro quality. An intermediate variable between micro and macro quality is needed to link them together, and this intermediate variable is named as medium quality. The connotation of medium quality must be able to explain the logical relationship between micro quality and macro quality, thus the quality concept of ISO is extended from a new perspective, so as to construct a scientific outlook on great quality. To this end, the paper puts forward the concept of quaternary quality, and constitutes the connotation of medium quality in accordance. It can link micro quality with macro quality organically, and make up the scientific outlook on great quality simultaneously. (As shown in fig. 1)

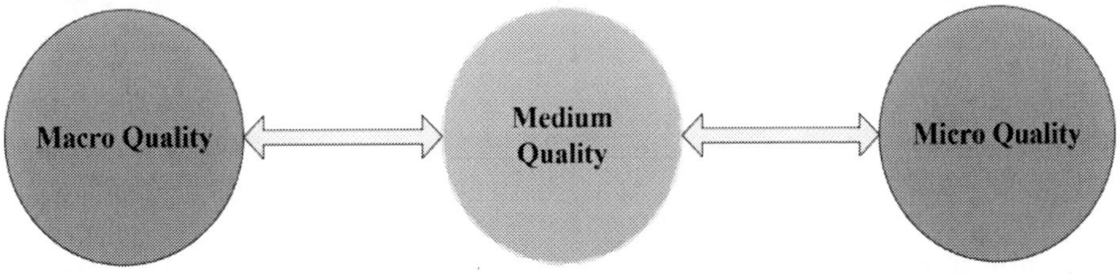

Fig. 1. The framework of the outlook on great quality

2 Literature review

Medium quality is explained by quaternary quality that can be described from four dimensions, that is, zero quality, unary quality, binary quality and ternary quality. Zero quality can be named as moral quality; unary quality also be referred to conformity quality; binary quality can be referred to applicability quality; ternary quality also be named as excellence quality. If regard the medium quality as a quality equation, the equation has four variables, that is to say, the equation is a quaternion equation which consists of zero quality, unary quality, binary quality and ternary quality.

2.1 The Connotation of Zero Quality

Zero quality, which is also named as moral quality, refers to the degree that products meet social ethics and public morality. Zero quality actually reflects the level of "quality faith" of enterprises. The integration of the levels of zero quality in all enterprises reflects the situation of "quality faith" in a region or a country. Some poor zero quality problems have turned up both in domestic and abroad frequently. At the beginning of the last century, Chicago slaughterhouse made sausage with dead pork, moldy pork and poisoned mice, and sold them both in domestic and international market. The incident not only brought about an extreme panic for domestic consumers, but also made international market give rise to an all-round blockade to US food, causing a half-decadent situation for US livestock. Due to the serious zero quality problems, the US food production caused country's credit crisis [4]. Another case is the Toyota recalls. The global outbreak of Toyota recalls in early 2010 directly exposed zero quality problems. Due to the devotion of exclusive attention to output, Toyota Motor Corporation ignored its quality and turned up some serious quality problems, such as power steering failure and engine failure, causing many injuries and deaths of people, as well as property damages. Because of the poor zero quality problems, Toyota had been recalled nearly 8.5 million vehicles, which made the credit of Toyota suffer an unprecedented blow, and also made a negative impact on Japan's economy[5]. For reducing the cost, China Shijiazhuang Sanlu Group co., LTD., once a famous Chinese dairy products enterprise, added industrial "melamine" into the infant milk powder to improve the protein content, causing serious kidneys damage of children who had ate it. The enterprise had much poor zero quality problems, and eventually made itself bankruptcy.

The quality culture undertook by enterprises and societies which take the values as the core and the moral spirit as the soul is the base to boost the level of zero quality. Carrying out all kinds of corporate social responsibility standards and certifications are the most important means for enterprises to enhance the level of zero quality. Improving the quality regulations of government is the external security to increase the level of zero quality. The enterprises that have zero quality problems have survival danger, so zero quality is their life. Enterprises whose products and business are closely related to people's life and property, animals and plants' health, should pay more attention to zero quality, and should focus their attention on the construction of quality culture and risk prevention mechanism, and also the education of risk consciousness.

2.2 The Connotation of Unary Quality

Unary quality, which is also named as conformity quality, refers to the degree that products comply with relevant quality standards. In order to judge the degree of meeting the requirements of standards, enterprises shall organize production based on the relevant standards, and products should be tested by the related inspection instruments. Unary quality is a kind of objective quality. Philip B. Crosby defined quality as "conformity quality", and believed that quality was tangible and also could be measured[6]. The level of unary quality actually reflects the level of standards, that is to say, high level of standards maybe achieve high level of unary quality. Unary quality usually be described with conformity and no conformity. If the products produced by enterprises and commodities circulated in the market have certification marks, it proves that the products and commodities conform to the relevant standards. Most enterprises have quality inspection departments to be in charge of testing the unary quality of their products. Both of the supervision on product quality and the testing on imported goods are the methods to judge the level of unary quality by governments. The product certifications engaged by various quality certification bodies are the documentary evidences of products to meet the unary quality. The CCC (China Compulsory Certification) mark printed on the product or its package means the product meets the safety standards. Up to March 2012, the number of CCC certification has grown to 22 major categories and 163 kinds of products, covering 10 industries that related to human health and safety, and have important influence on national economy. Nearly 618,000 CCC certificates licenses had been issued to 88,000 enterprises [7].That is to say, nearly 618,000 kinds of products had met the requirement of unary quality.

Quality inspection is one of the main methods for unary quality management. Furthermore, standards, testing equipment and laboratory facilities are the most important means to improve the unary quality. The level of the technical standards determines the level of unary quality directly, and testing level determines the reliable level of testing results for unary quality. The main method to improve the testing level is standardization. For example, implementing ISO 10012-1 and ISO 10012-2 standards can assure the measuring reliability of inspection departments of enterprises. To engage in product quality certification, testing and calibration laboratories must carry on ISO/IEC 17025 standard, and gain the laboratory accreditation. Sampling technology is also an important tool to assure the unary quality. Most of the quality inspections are sampling inspection. The rationality of sampling schemes determines the reliability of unary quality directly. There are many sampling standards for sampling inspection. For the enterprises that produce spare parts and semi-finished products, outsourcing services, or the enterprises that produce products based on order forms, or the enterprises that are in the middle and lower positions of the supply chain, unary quality is their life, and conforming to the standards is their highest goals.

2.3 The Connotation of Binary Quality

Binary quality, which is also named as applicability quality, refers to the degree that products meet the requirements of customers or market. When the product standards are not adapted to the market requirements, even if the product quality achieves the high level of unary quality, the products also have no market share and no benefits, because the product quality doesn't meet the requirements of customers. While meeting the requirements of the standards, many products are surplus and overstocked because of the dissatisfying with market requirements. For instance, Global Satellite Mobile Communication System, which was invented by American Iridium Satellite Company, was unable to form the stable customers bases because of its quality problems, such as terminal bulky, capacity insufficient, communication prolonged, lower spectrum utilization, and the failure to provide the personal mobile communication business based on the handsets. Furthermore, its standards lag behind the requirements of the market[8], so the enterprise officially declared bankruptcy on March 17, 1999. For another instance, the mobile phones produced by Nokia and Ericsson companies once became the main supply products in the market because of using the standard of second generation mobile communication. But when the smart-phone appeared, they lost the mobile phone market and were replaced by Apple and Samsung. Nokia and Ericsson companies failed to create applicable products in the light of new standards, so their product quality stayed in the original unary quality and couldn't satisfy the new requirements of customers. Therefore, improving the level of binary quality and satisfying the requirements of customers are the foundation of sustainable development of the enterprises. There are lots of methods and means to improve the level of binary quality. Market research is a market-oriented way to improve the level of binary quality. It can find the actual and potential requirements of the customers. Then new standards are set based on these requirements, and thus the products that can satisfy these requirements are produced in terms of these standards. Innovation creates the market or drives the requirements, and these requirements are led by innovation achievements. The achievements of science and technology innovation create customers' requirements, which are the unexpected requirements for customers. Apple Inc. innovates the smart-phones by applying breakthrough technology, which created the market directly and brought the unexpected requirements for customers, so as to improve the level of binary quality. The construction of quality culture and the management of customers relationship which regard "Customers are always right" as the core principle are the effective methods to improve the binary quality of enterprises. The standard of ISO/TS10004:2010 quality management customer satisfaction monitor and measure guide is also an important mean to improve the binary quality. For the enterprises whose products sale to the terminal market directly or direct-to-consumer, binary quality is their life. Enhancing the level of binary quality is the core factor of quality managements for these enterprises.

2.4 The Connotation of Ternary Quality

Ternary quality, which is also named as excellence quality, refers to the degree that enterprises pursue the business excellence. The essence of ternary quality is the business quality of enterprises and the level of brand construction. Enterprises gain excess profits by low cost of investment and excellent management, so as to bring great value to customers. On the contrary, if enterprises only stay in binary quality level, they may be unable to obtain excess profits. There are many enterprises that may have good shares in the market,

 https://doi.org/10.1051/shsconf/20184901002

but their benefits are low, and even have no benefits. For example, the ties produced by enterprises that are located in Shengzhou county, Zhejiang province, had shared 60% of the international market, but the profits were less than 1%. Because these products failed to become an international brand, although the level of binary quality of the enterprises that produced these products was very high, their benefits were low[9]. For another example, the profit rate of Volkswagen in China was 11.3%, but the rate of domestic automobiles was merely less than 3%[10]. The reason is very simple: Volkswagen is a famous international brand, but our car manufacturers are not. Quality is the core factor of building a brand, and the path of building a brand is the pursuit of excellence business to enterprises. There are many methods to improve the ternary quality, such as marketing, advertising etc. It turns out that the Six Sigma management and Criteria for Performance Excellence are ones of the most effective and widely used methods for enterprises to improve ternary quality, pursue excellence business, and build a brand. No matter what kinds of enterprises, only if they have the pursuit of lasting forever and sustainable development mode, the high level of ternary quality is their essential foundation.

3 Logical Relationships

Unary quality, binary quality and ternary quality have a progressive logical relationship, that is, if enterprises want to pursue ternary quality, they must satisfy the requirements of unary quality and binary quality simultaneously. Zero quality is the foundation of all the other three qualities. When zero quality is poor, the quality tower will collapse, and everything will return to zero. Therefore, the structure of medium quality is a pyramid, which can also be named as quality pyramid. (As shown in fig. 2)

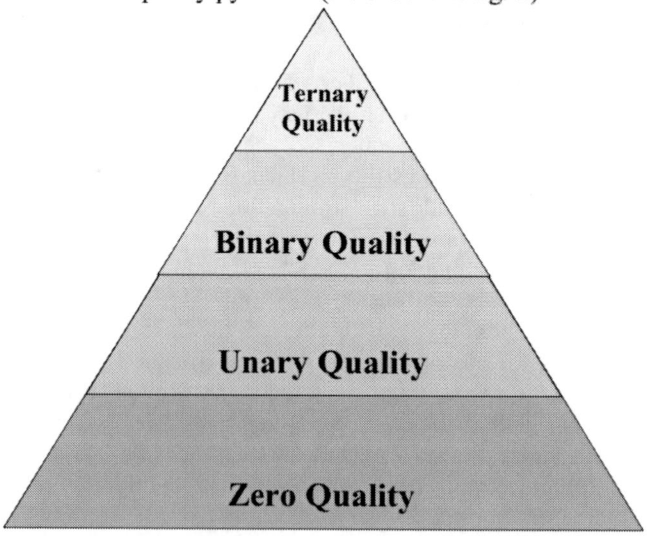

Fig. 2. Quality pyramid.

Quality pyramid links micro quality and macro quality together. The core elements of macro quality are credit, voice in standards, market share and the number of international brands. For a country or a region, the level of credit depends on the quality faith, and zero quality is the foundation of building quality faith. Voice in standards, which mainly refers to the voice in the international market, represents the influence on the country or region. If the standards are controlled, then the market will be controlled. The high level of unary quality is the most important way to promote the standardization construction. Only high standards can bring about high quality, and the high standards depend on a high level of

https://doi.org/10.1051/shsconf/20184901002

scientific and technological innovation, while the technological innovation can drive economic development. The international market share is closely related to the international market competitiveness, and only the pursuit of a high level of binary quality can lead to a high level of market share. The amounts of international brands affect the core competitiveness of a country or a region directly, and also determine their development level. For the annually top 500 of international brands ranked by World Brand Laboratory, the number of brands in United States, Japan, Britain, France and Germany have been always top five[11]. These five countries also have been the highest level of development quality in the world. Therefore, the pursuit of excellence business and building brands are ones of the most effective means to improve the ternary quality.

4 The Outlook on Great Quality

Medium quality, which makes quaternary quality as the core factor, combines micro quality with macro quality naturally and organically, then the outlook on great quality can be formed accordingly. The outlook on great quality shows that the development quality of economy in a country or a region depends on the quality of products. Only promote the level of product quality down-to-earth can we improve the development quality of economy, and the economic development processes of United States, Japan and Germany have proved this regular pattern. These three countries have the highest level of macro quality in the world, because they have the highest level of micro quality. Their micro qualities also have distinguishing features: the quality feature of United States is innovation; the quality feature of Japan is lean and the quality feature of Germany is reliability. Therefore, the outlook on great quality shows that only a high level of product quality can assure the development quality of national or regional economy. Improving the level of quality management of the enterprises and the level of quality governance of a society are the foundation to enhance the development quality of economy.

References

1. C. Hong. Journal of Wuhan University,129-134(2010)
2. Lang Zhizheng. "The Outlook on Great Quality That I Had Seen." Supervision and Selection (November): 4(2002)
3. S. Mingshun. Quality Management, second edition. Beijing: Science Press (2012).
4. U. Sinclair. The Jungle. Si Chuan: Sichuan Literature and Art Press (2010).
5. X. Guojun."'Toyota Recalls' is a Game about what?" China equipment (April): 90 (2010).
6. P. Crosby. Quality Is Free The Art of Making Quality Certain. Beijing: China Renmin University Press (2006).
7. S. Jingjing."Perfecting Compulsory Product Certification System in China." (April): 14(2012)
8. L. Lichun. Telecommunications Science, **31**(1999).
9. L. Yabiao and S. Yiying. "Shengzhou Tie : Aphasia of The World's Largest." Economic Information Daily (2010)
10. Fortune. The World Fortune Rankings of Top 500 in 2013 [EB/OL] http://money.cnn.com/magazines/fortune/rankings/?iid=F_Sub(2013).

https://doi.org/10.1051/shsconf/20184901002

11. World Brand Laboratory. The World Brand Rankings of Top 500 in 2012 [EB/OL] : http://www.worldbrandlab.com/world/2012/

The benefit attributes of the Indonesian National Standard (SNI) product

Muhammad Rosiawan[1,*], *Moses* Laksono Singgih[2], and *Erwin* Widodo[2]

[1] Department of Industrial Engineering, Universitas Surabaya, Jl. Raya Kalirungkut, Surabaya, Indonesia

[2] Department of Industrial Engineering, Institut Teknologi Sepuluh Nopember, Jl. Arif Rahman Hakim, Surabaya, Indonesia

Abstract. This research aims to measure the benefit of the Indonesian National Standard (SNI) product not only came from the company perceptions but also came from the government, academician, and consumer perceptions in Surabaya. The empirical study result showed that all of the respondents had a positive perception on the SNI product. They stated agreement that the SNI product had benefit on the attributes: quality guarantee of product, safe to use, a fair prices, quality of raw materials, and efficiency. Those attributes support the enhancement of company's competitiveness. However, most of the respondents of companies and consumers state did not know about the quality guarantee of the SNI product. The results were contradictory, the respondents who came from the companies knew more about quality guarantee of the SNI product compared to the other respondents. Furthermore, the respondents of academicians, consumers, and particularly companies state that they did not know about the easiness of the SNI certification process and the government provide an incentive for the registration fee of the SNI certification. This ignorance will impact on the perceptions that process of the SNI certification was complicated and costly. Besides, it showed that the government was less to conduct socialization for the certification program compared to socialization for the standard program. Keywords: *Attributes of SNI product, Certification, Competitiveness, Respondent's perceptions*

1 Background

A company that implements a management system standard intended to provide information (such as production efficiency, the quality of raw materials, quality product, and safety for use) to companies and customers that the resulting product comes from a controlled management system and always performs continuously improvement [1]. Unfortunately, this kind of information is often unknown to other stakeholders and become only internal information of the company. Therefore, standard and certification implementation programs in company help to disclosure of information about the hidden attributes to suppliers, customers, and other stakeholders.

* Corresponding author: mrosiawan@staff.ubaya.ac.id

Most of the empirical study discusses the benefits attributes of the ISO/SNI standards mainly on the operational and financial company's performance. The primary data collected from filling of questionnaires by respondents and the secondary data obtained from the public data. The research findings showed:

a) The company got less benefited if the certification is the customer demands only [2]. But the company which has ISO 9001 certificate (caused by internal motives) to get positive benefits on operational and financial performance [3],
b) Companies benefit from operational and financial performance. Kinds of operational performance are improvements in product and service quality (Psomas & Pantouvakis, 2015), company efficiency [4], to provide product safety, employee safety and health as well as the environment [5], improvements in the relationship between process management & quality improvement, process management & quality support tools [6], improved manufacturing parameters [7], technology management, participation in quality improvement programs and quality control techniques [8]. While kinds of the financial performance are increased sales revenue and return on asset [9]. But the benefits of this financial performance depend on the seriousness of the company in implementing the standards [10],
c) The SNI products strengthen the domestic market and competitiveness of the export products [11-12]

Other activities in line with standardization are conformity assessment as a statement that products, processes, systems, persons or institutions have met certain requirements, which may include testing, inspection, certification and accreditation activities [13]. Benefits of certification by credible third parties include a) reducing the frequency of product failures in the market [14], b) reducing asymmetric information between sellers and buyers about product quality [15], and c) enhancing quality and image of the company [16]. Furthermore, an empirical study in Hong Kong shows that 75% of companies surveyed stated that the cost of certification is inexpensive when compared to the benefits obtained [2].

From those studies, most of the research object was the companies. While the object of research involving consumers, academicians and government was still limited, such as the study of [5] which discussed the perception of the consumers in four cities in Indonesia (Denpasar, Banjarmasin, Mataram and Manado) as respondents related to attributes of the SNI products such as product safety, safety, and health as well as environmental sustainability. Respondents of study to review of the SNI role include producers and exporters, government agencies, and employers' associations in production centers and export locations in five regions in Indonesia (Jakarta, Jawa Barat, Jawa Timur, Sumatera Utara, and Banten) [11]. Furthermore, respondents of the study of public perception on the SNI product consist of academics, companies, consumers, and government in ten big cities - Indonesia (Medan, Riau, Jakarta, Bandung, Yogyakarta, Surabaya, Samarinda, Makassar, Manado and Bali) [12]. Although the research has addressed several attributes of product SNI from the perceptions of various respondents, these studies were still partial. These studies did not address other attributes of SNI products and then arrange these attributes in a structured way to form an image of SNI product.

Therefore, this paper discusses about the respondents perception on attributes of SNI product such as raw material quality, efficiency of process, quality of product, safe to use, a fair price, product or company competitiveness, easiness of certification process, and incentives for registration fee of SNI certification. In addition, the study analyze the relationship between type of occupation with attributes of the SNI product and certification.

In order to extent those research, this paper discusses the perceptions of respondents (consumers, academicians, companies, and government agencies) on the benefit attributes of the SNI product. The attributes that exist in the SNI product for the design of the questionnaire are structured as follows (Figure 1.)

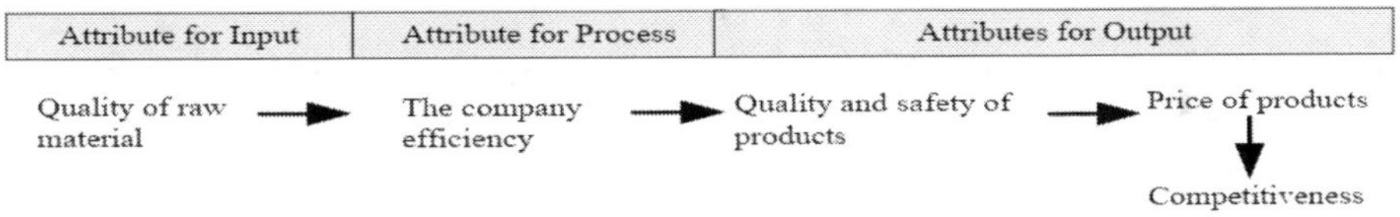

Figure 1. Relationship between each attributes of the SNI product

Figure 1 shows that the quality of raw material has relationship with the company efficiency such as efficiency of production process. The company efficiency has relationship with quality and safety of the product. Furthermore, the fair price-high quality has relationship with the company or product competitiveness. Therefore, design of questionnaire based on relationships between these attributes in order to measure respondents' perceptions of the benefit attributes of standard (SNI) implementation.

The paper structured as follows. Firstly, the background chapter that discusses the implementation of standards and its impact. Secondly, explains the empirical research method used. Thirdly, discusses the key findings. And lastly, the conclusion, limitation and opportunities for next research.

2 Methodology

2.1 Selection criteria of respondents

Four hundred respondents in Surabaya is divided into four groups with a sample proportion: 30% for consumers, 30% for companies, 20% for academicians, and 20% for government agencies. Respondents were randomly selected and have knowledge of SNI and certification.

2.2 Design of Questionnaire

The questionnaire was designed to follow the attributes in Figure 1. Respondents were asked to show their agreement for each question on a four-point scale where 1 (Strongly disagree), 2 (Disagree), 3 (Agree), and 4 (Strongly agree). If the respondents did not know about the questions, then respondents are able choose the answer 0 (I do not know). In addition, the questionnaire also asked respondents regarding their perceptions of the certification process and incentives for the registration fee of certification. Table 1 shows the profile of respondents.

Table 1. Type of occupation of respondents

Field of occupation	Type of occupation	Frequency	Percent
Government Agencies	Army/Policeman	18	4.5
	Staff government	61	15.3
Academicians	Lecturer	24	6.0
	Teacher	59	14.8
Companies	SME Employee	38	9.5
	SME owner	40	10.0
	State-owned enterprises staff	47	11.8
Consumers	Housewife, medical doctor/dentist, sales, staff in church office, entrepreneur, lawyer, cleaning service	113	28.3

3 Key Findings and Discussion

From Table 2, approximately 70% of the respondents agree and strongly agree for attributes of the SNI product are high quality, safe to use, fair price, quality guaranteed of raw materials, improve efficiency and competitiveness. However, the perception of respondents who agree and strongly agree tends to decrease (about 50%) for attributes about easiness of the certification process and the government provided an incentive for registration fee of SNI certification. Most of the respondents did not know the attributes about the certification.

Table 2. Perceptions of Respondents on Benefit Attributes of SNI product

Attributes		Percent				
		0	1	2	3	4
a.	Quality guarantee of raw materials	25.8	0.0	2.3	56.0	16.0
b.	Improves the company efficiency	11.8	0.0	4.3	71.3	12.8
c.	Quality guarantee of product	11.3	0.3	9.0	62.0	17.5
d.	Safe to use	10.8	0.0	5.5	64.8	19.0
e.	Have a fair price	18.6	1.0	10.0	59.0	11.5
f.	Easiness of certification process	28.5	5.8	13.3	41.5	11.0
g.	An incentive for registration fee of the SNI certification	32.0	0.8	14.5	43.5	9.3
h.	Improves product and company competitiveness	17.3	0.5	4.3	65.0	13.0

The result dependency relationship between types of occupation with attributes of SNI product used Pearson Chi-Square Test (Table 3). Table 3 shows that there is dependency relationship (whereas the value of Sig. (2-sided) is less than 0.05) between the types of occupation with the perceptions for: a) quality guaranteed of the SNI product, b) easiness of the SNI certification process, and c) the government provided an incentive for registration fee of the SNI certification.

Table 3. Result of dependency test (Pearson Chi-Square Test)

Attributes		Type of occupation
a.	Quality guarantee of raw materials	0.416
b.	Improves the company efficiency	0.908
c.	Quality guarantee of product	0.005*)
d.	Safe to use	0.204
e.	Have a fair price	0.146
f.	Easiness of certification process	0.022*)
g.	An incentive for registration fee of the SNI certification	0.002*)
h.	Improves product and company competitiveness	0.279

*) Statistically significant when the value of Sig. (2-sided) is less than 0.05

Furthermore, Table 4 is used cross-tabulation analysis to determine the control variables (types of occupation) which contributes to its dependency relationships. Of Table 4 shows that the respondents working in the company have higher percentage about ignorance of quality guarantee of the SNI product compared to the respondents working in the government agencies or as an academician. The result is contradictory because the respondents who working in the company are directly involved in the production process and quality of product and they should know more about the quality product compared to other respondents. Furthermore, many of respondents (academician, consumers, companies and government agencies) state that they also did not know about the certification attributes (e.g. the easiness of the SNI certification process and an incentive for the registration fee of the SNI certification) compared to the SNI product attributes. The result state that the socialization of the certification program is rare.

Table 4. Cross-tabulation between the type of occupation and attributes of the SNI product

Attributes	Level of perception	Types of occupation (Percentages)			
		Government Agencies	Academicians	Companies	Consumers
Quality guarantee of product	0	0.5	0.0	4.3	4.0
	1	0.0	2.5	0.3	0.0
	2	3.0	2.5	4.3	2.0
	3	16.0	13.3	16.0	16.8
	4	3.0	2.5	6.5	5.5
Easiness of certification process	0	3.0	6.0	9.5	10.0
	1	0.5	1.3	2.8	1.3
	2	3.0	3.8	2.5	4.0
	3	11.5	8.0	12.5	9.5
	4	1.8	1.8	4.0	3.5
Easiness of certification process	0	3.0	7.8	10.3	11.0
	1	0.0	0.0	0.8	0.0
	2	3.3	3.5	4.5	3.3
	3	12.5	7.8	11.5	11.8
	4	1.0	1.8	4.3	2.3

4 Conclusion

The results show that the products certified by SNI already have attributes such as quality of raw material, efficiency of the company, product quality, safety in use, and fair price. Most of respondents agreed that the SNI product is able increase the competitiveness.

The research findings that different from previous studies because of the asymmetric information for the SNI certification program. Most of the staffs and owners of company who did not know the information about the certification process and an incentive for registration fee SNI certification provided by the government. This ignorance may cause the SNI certification process is complicated and costly. While the companies strongly agreed that the application of standards and certifications improve the company efficiency. It should be the concern of government agencies related to standards and certification to increase the frequency socialization about certification programs to companies. The frequency socialization is able to maintain the positive perceptions of the implementation of standards and the certification programs.

This research is still limited to respondents in Surabaya. Therefore, it is necessary to expand the scope of its respondents in Indonesia to obtain more general results. Furthermore, the future research is able to measure the impact of standards and certifications on the company performance through using the company's operational data.

Acknowledgments

The research was funded by Deputy of Information and Socialization of Standardization - National Standardization Agency for the data collection. Besides, the research conduct by students and Head of Quality Performance Management Laboratory– Department of

Industrial Engineering, University of Surabaya are as surveyors and a coordinator of survey.

References

1. E. L. Psomas and A. Pantouvakis., *TQM J.*, **27**, 5, 519–531 (2015).
2. H. K. N. Leung, K. C. C. Chan, and T. Y. Lee, *Int. J. Qual. Reliab. Manag.*, **16**, 7, 675–691 (1999).
3. P. Sampaio, P. Saraiva, and A. G. Rodrigues, *Int. J. Qual. Reliab. Manag.*, **28**, 9, 929–950 (2011).
4. D. Tzelepis, K. Tsekouras, D. Skuras, and E. Dimara, *Int. J. Oper. Prod. Manag.*, 26, 10, 1146–1165 (2006).
5. F. Isharyadi, A. Wibowo, and Suminto, *J. Stand.*, **19**, 1, 71–78 (2017).
6. E. L. Psomas, C. V. Fotopoulos, and D. P. Kafetzopoulos, *Bus. Process Manag. J.*, **17**, 3, 437–460 (2011).
7. T. Koc, *J. Mater. Process. Technol.*, **186**, 207–213 (2007).
8. A. Gupta, *Ind. Manag. Data Syst.*, **100**, 9, 451–455 (2000).
9. F. Starke, R. V. Eunni, N. M. M. D. Fouto, and C. F. de Angelo, *Manag. Res. Rev.*, **35**, 10, 974–997 (2012).
10. P. Sampaio, P. Saraiva, and A. Monteiro, *Int. J. Qual. Reliab. Manag.*, **29**, 8, 891–914, (2012).
11. BPPKP-Kemendag, "Kajian Peranan SNI Untuk Penguatan Pasar Dalam Negeri dan Daya Saing Produk Ekspor Pusat Kebijakan Perdagangan Dalam Negeri Badan Pengkajian dan Pengembangan Kebijakan Perdagangan Kementerian Perdagangan Republik Indonesia," 2015.
12. Kedeputian-IPS-BSN, "Laporan Akuntabilitas Kinerja Instansi Pemerintah Kedeputian Informasi dan Pemasyarakatan Standardisasi Tahun Anggaran 2016," 2017.
13. M. Eto, M. Song, J. Henry, H. Lee, H.-D. Yang, and H. Lee, *Standardization: Fundamentals, Impact and Business Strategy*. APEC Secretariat, 2010.
14. J. A. Clougherty and M. Grajek, *Int. J. Ind. Organ.*, **36**, 70–82 (2014).
15. I. Vertinsky and D. Zhou, *Int. Mark. Rev.*, **17**, 3, 231–252 (2000).
16. B. Buehler and F. Schuett, *Eur. Econ. Rev.*, **70**, 493–511 (2014).

Three activities to promote the development of standard professionals

Manabu Eto[1,*]

[1]Institute of Innovation Research, Hitotsubashi University, 2-1, Naka Kunitachi Tokyo Japan

Abstract. I was involved in three projects to educate standard professionals in 2017. The first was conducted for first through third-year university students as a part of standardization education in collaboration with several universities in the west Tokyo area, and many students participated. In the second, the Japan Standards Association began a system for registering standard professionals, and prepared and implemented a training system for acquiring registration qualifications. In the third, Yamaguchi University used government funds to create education materials for instructors for standardization education. These three projects are aimed at solving the problems found in past standard professionals' education. I will report the details of these three projects as a case example of standard professionals' development activities in Japan.

1 Introduction

Efforts to educate and train standardization-related professionals have a long history, but Japan only started undertaking such efforts at a policy level in 2004. In that year, the Koizumi Cabinet formulated a set of guidelines titled the "Big-Boned Policy" (honebuto no hōshin), which highlighted the "strategic acquisition of international standards." In the same year, the government released the "Plan for Promoting Intellectual Property," which underscored the importance of bolstering strategic efforts related to international standardization.

In tandem with these policy developments, METI released in June 2004 the "Action Plan for Bolstering the Base for International Standardization Practices" and, in 2006, the "Strategic Goals for International Standardization." These guidelines stipulated two key goals: (1) Submit twice as many proposals for international standards than in the passing year, and (2) be the "core country" (the country that holds an advantageous position in the formulation of a given international standard) in as many standards proposals as Western nations. The guidelines also outlined a concrete measure for achieving these goals: "training standards experts who are internationally fluent." Tasked with examining specific strategies for how to implement this measure, the Japanese Industrial Standards Committee (JISC) established a special committee on training and education. On July 16, 2008, the committee then released its report, which became the first report in Japan specializing in

* Corresponding author: eto-manabu@iir.hit-u.ac.jp

training standardization experts. The report was titled "Future Approaches for Training Standardization-related Professionals."

Following the publication of the report, standards-related organizations in Japan undertook various initiatives; JISC (an ISO and IEC member) and the Telecommunication Technology Committee (TTC; an ITU-T member) provided textbooks on standardization [1][2], and the Japanese Standards Agency (JSA) likewise prepared educational materials and published a textbook [3][4].

In January 2017, JISC's "Working Group for Training Standardization-related Professionals under the Public-Private Strategic Council for Standardization" formulated the "Threefold Action Plan for Training Standardization-related Professionals." The Action Plan first defined three categories of standardization-related professionals: (1) managerial professionals responsible for developing rule-making strategies, (2) standardization experts, and (3) professionals who support companies' standardization efforts. The document also outlined three actions that companies should immediately take in order to train such standardization-related professionals: (1) improve management's awareness about standardization, (2) clarify organizational frameworks and the system of personnel evaluation, and (3) formulate and implement a plan to train the necessary professionals.

In response to these recommendations, we commenced three attempts to train up new standardization-related professionals. I will introduce these three attempts below.

2 Joint Lecture Course in Multiple Universities

As regards the third category of standardization-related professionals, the Threefold Action Plan, focusing on professionals who support standardization efforts, emphasizes the important role that universities (at both the undergraduate and graduate level) play in this. According to a report of the last ICES[5], Japanese universities are not providing adequate education in standardization, either in terms of quantity or quality. The main reason why standardization education is not proliferating among universities is the shortfall in instructors who are capable of educating people in standardization. So far, I have held lecture courses on standardization in more than ten universities, which likely covers 30–40% of the number of standardization experts training currently implemented at universities. However, there is a limit to how much time one person can spare for such lecture courses.

Therefore, the working group's leader, Haruo Takeda (Hitachi Ltd.), organized a joint lecture program on standardization involving five national universities in Tokyo's Tama area (Tokyo University of Foreign Studies, Tokyo University of Agriculture and Technology, the University of Electro-Communications, Tokyo Gakugei University, and Hitotsubashi University).

The lecture program consisted of a week-long intensive course held in September 2017 in Tokyo University of Foreign Studies. I was responsible for the first half of the course (approximately 10 hours), which focused on basic knowledge about standardization, managing intellectual property, and business. For the latter portion, experts from nine entities among industrial sector and government agencies gave detailed lectures on the trends in the relevant field regarding standardization and international negotiations.

The key point of the course was to proactively convey information that closely linked standardization and rule-making. To this end, we put the following message at the forefront: you are either bound by the rules, or you make the rules. A lecture meeting to advertise the course was organized in the late spring of 2017 by a core industry team comprising Hiroaki Nakanishi (chairperson of Hitachi Ltd.), Hiroshi Tomono (Nippon Steel & Sumitomo Metal Corporation), and Tamotsu Nomaguchi (advisor and former president of Mitsubishi Electric).

Since the actual course was held in Tokyo University of Foreign Studies, 70% of the attendees were studies from the said university, while the remaining 30% were students from other universities. Around 100 students attended the course. Before attending the course, most of the respondents said that they knew hardly anything about standardization. However, during the course, they learned not only about standards themselves but also about who formulates standards and how. The course was well received by the students; 70–80% of them expressed satisfaction, and many reported that they were pleased to have attended. We plan to hold the course also in 2018.

3 Systems for Training and Licensing Experts in Standards Development

The 2008 JISC report stated that establishing a licensing system, such as a national exam, would be an important step in training standardization-related professionals. After subsequent deliberations, however, JISC abandoned the idea, and since then, it has never taken any action to license standards experts or expand the fluidity thereof.

In 2013, JSA implemented the "Program for Developing a Network of Experts in International Standardization." Under this program, JSA created a registry database of Japanese experts in international standardization, and it launched a system under which companies who require the services of international standardization experts could apply through JSA to hire a registered expert. However, it was up to the experts themselves whether they would apply to register to the JSA database, and JSA regulated the skillset in its database by setting certain criteria, such as requiring applicants to have attended international conferences in the past. For these reasons, JSA failed to get an adequate number of experts registered, and the system was not widely used by companies.

Therefore, JSA decided that it would need to launch a new registry of standardization experts, and to this end, it deliberated on what kind of training system it would need to ensure that experts had the requisite competences. I participated in these deliberations as chair of the investigative committee.

Based on the findings of the investigative committee, JSA developed two training courses, and it stipulated that individuals could register themselves as standards development experts if they have attended the training and have two years of on-the-job practical experience.

To describe the training in more detail, trainees must attend a 1-day (6-hour long) course titled "basic course in strategically applying standardization" and pass an end-of-course exam. At present, I am in charge of this 6-hour course, and as the chair of the editing committee, I use a JSA textbook on standardization [3]. I also designed the end-of-course exam. We asked to attendees of the second course (held in Tokyo) and third course (held in Osaka) stated their purpose for attending. Few of the respondents said that they attended the basic course in order to register with JSA; the majority attended either to apply the knowledge in business or to boost their own knowledge.

Those who pass the end-of-course exam are able to attend a more advanced course, titled "standardization expert course." This course lasts for four days, and I am in charge of the first day. As for the remaining three days, officials from ISO, IEC, JIS (especially those who contributed their texts to the abovementioned textbooks and also), as well as in-house standards experts, deliver lectures about how standards are drawn up in their particular fields.

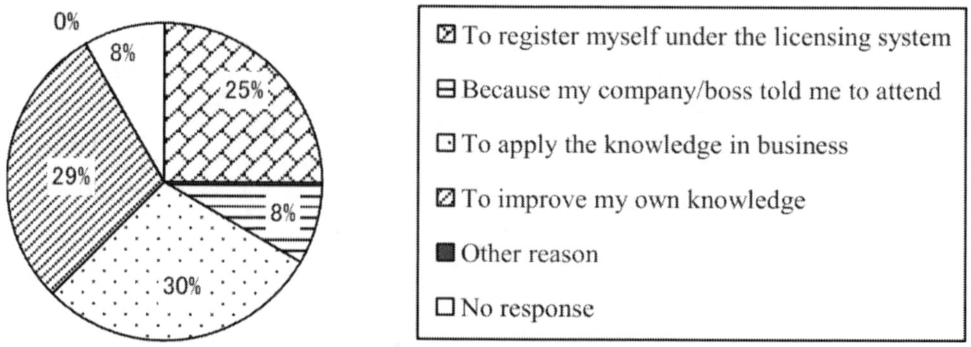

Fig.1: Reasons for attending the standardization expert course in strategically applying internationalization.

Those who attend the course and pass the end-of-course exam can register themselves on JSA's database as provisional standardization experts. Then, once they have gained two years of operational experience (participating in an official capacity in international or national standardization processes) since the time of their registration, they can register as full standardization experts.

Figure 1 shows the results of questionnaires in which the attendees of the standardization expert course stated their purpose for attending. Participation for qualification registration is also small in this course. However, 80% of the attendance registered as experts after completing the course.

Those registered can renew their license every two years, but the renewal process costs money. As such, only those who are willing to apply their expertise will register. Nevertheless, almost one year since the system's launch, as many as 48 people are registered, demonstrating the potential for getting more people registered as the system spreads widely.

As a special dispensation for the period up to March 31, 2019, those with more than three years' experience in standardization practice are permitted to register as standardization experts without having to attend the above training. This dispensation applies to the system's initial launch period, but given that there are many experts in Japan who already have sufficient knowhow and thus do not require training, this dispensation is arguably an essential measure to bring these experts on board as registered standards experts.

Incidentally, in the year 2018, we held the "basic course in strategically applying standardization" three times and the "standardization expert course" twice.

4 Developing Materials for Universities

As mentioned above, a number of educational materials have already been developed. I myself have referred to a number of these, but considering the shortage of instructors who are competent to educate in standardization, simply issuing textbooks or lecture materials will not be enough to expand standardization education in universities. Therefore, in 2017, METI issued funding for a project to develop lecture notes for instructors. The notes were to be developed by integrating the materials I used in my lectures to university students with the content of these lectures.

The people who took charge of the project were Professor Tomohisa Kimura and Associate Professor Masato Sasaki, from the Yamaguchi University's Intellectual Property Center. Yamaguchi University selected Kimura and Sasaki because its Intellectual Property Center had prepared lecture materials on intellectual property the year before, and so they could use these materials to help standardize the knowledge in the lecture notes.

As part of the project, I conducted a 2-day long intensive lecture course at Yamaguchi University, videoing the entire proceedings. A text transcript was prepared and combined with the slides I used, thus creating something akin to a lecture transcript.

Yamaguchi University edited the video footage, preparing a 10-hour long lecture video. In the transcript, my verbatim speech was extensively edited into a format that could be read out as is by instructors in standardized lectures. This process culminated in an approximately 300-page instruction manual. Videoing a lecture can enable e-learning, but our purpose was to ensure that instructors at each university could deliver standardized lectures themselves; therefore, we wanted instructors to deliver the lectures using the textual material in the manual, and to use the video footage only for their reference.

For 2018, we plan to conduct an awareness raising campaign aimed at promoting the use of the lecture manual and video in many universities across Japan and encouraging each university to develop their own lectures based on the material. Arrangements have already being made to introduce the material at the assembly of national universities.

5 Conclusion

As I presented at last year's ICES2017 and APEC-IEC-ISO workshop [6], the following is a summary of the issues recognized in standardization education in Japan.

a. There are no suitable textbooks

b. There are too few personnel capable to carrying out training

c. Students and business people have no interest in standardization (the merits of learning standardization are not apparent)

d. There are no places for working adults to learn about standardization.

e. Company management does not recognize the importance of standardization knowledge.

The three approaches in this study were attempts to resolve these issues. First, as mentioned in (1), regarding the issue of (a) textbooks, there has been some degree of improvement made in textbooks for standard professionals involved in standardization activities. In (2), running the lecture in collaboration with several universities, although it compensated for (b) the shortage of education personnel, it succeeded in (c) stirring up interest in standardization. For (3), the standard development expert system and the preparation of a training system for that, we created (c) merits for learning standards and were able to provide (d) a place for working adults to learn about standards. Then, in (4) the development of university education materials, it became possible to expand standardization education at universities by compensating for (b) the personnel shortage.

As mentioned above, I will continue the standardization lectures this academic year as well, in collaboration with several universities and will also hold training sessions for standard development expert systems. Furthermore, I am planning to develop and publish an inexpensive ($5) textbook on standardization for university department students, centered on Yamaguchi University.

In this way, the major remaining issue is the fact of (e) company management not recognizing the importance of standardization knowledge. To resolve this issue, I am working on increasing the opportunities for direct training for company management, but it is not easy for company managers to set aside time to attend training, and it is also not easy to change the minds of older company managers.

Perhaps this issue can only be dealt with by spending time. By increasing standardization education at universities and graduate schools, we can expect the awareness toward standardization to change greatly by the time the students who learned there become company management themselves. It is particularly necessary to increase standardization education not just in STEM education fields, but also in economics, management departments, and business schools. For such educational opportunities to happen, it may also be necessary to create standardization textbooks geared toward business activities.

As we have seen from the above, since JISC released its action plan in January 2017, there have been many undertakings in Japan to train standardization-related professionals. I have engaged in many of these efforts as an instructor. However, we have still not fully established a system in Japan for training standardization-related professionals. Looking forward, we need to continue making various efforts to expand standardization education.

References

1. M. Eto, (ed.) Hyōjunka jitsumu nyūmon (shisaku-ban) [Standardization practice for beginners <pilot edition>], Accreditation Unit, Industrial Science and Technology Policy and Environment Bureau, Ministry of Economy, Trade and Industry (2010)
2. Ministry of Internal Affairs and Communications, Jōhō tsūshin bunya ni okeru hyōjūnka katsudō no tame no hyōjunka kyōiku tekisuto [Standardization education text for standardization practices in information and communications], Telecommunication Technology Committee. (2013)
3. Yoshida et al., Hyōjunka kyōiku puroguramu kaihatsu kyōzai [Materials for standardization education programs], Japan Standards Agency (2009–2011)
4. M. Eto, (ed,) Hyōjunka kyōhon [Standardization textbook], Japan Standards Agency (2016)
5. M. Eto, Education of Standardization in Japan, ICES2017 (2017)
6. M. Eto, Human resources of Standardization in Japan, APEC-IEC-ISO workshop on Developing Competence Requirements and Career Path for Standards Professionals (2018)

Analysis of ISO 9001:2015 certification readiness of JP-Graha product of Jasaraharja Putera using gap analysis

Bambang Purwanggono[1*], *Arfan* Bakhtiar[1], *Rifky* Rahman H[1],

[1]Industrial Engineering Department Faculty of Engineering, Diponegoro University, Jl. Prof. Soedarto, SH, Undip Tembalang Campus, Semarang, Indonesia 50275

Abstract The fierce competition of global economy requires company to have a good Quality Management System (QMS). Efforts done by the company included applying standards of Quality Management System (QMS) ISO 9001. ISO 9001 is a standard that contains requirements of quality management system, published by the *International Organization for Standarization (IOS)*. The object of this research is JP-GRAHA PT Jasaraharja Putera. JP-GRAHA is one of the product from PT Jasaraharja dealing with fire insurance. JP-GRAHA already implemented ISO 9001:2008. But the implementation of ISO 9001:2008 suffer from a decline in consistency, meanwhile the company must prepare to *upgrade* into ISO 9001:2015. This research aims to identify the readiness of the company in the facing Certification of ISO 9001:2015 and provide recommendations to improve the readiness of the company. The method used in this research is Gap Analysis to evaluate current company state and assess the readiness of the company in facing certification of ISO 9001:2015. Gap Analysis is a checklist made based on the existing requirements icluded in ISO 9001:2015. The results from the calculation using the checklist will show how the readiness of JP-GRAHA PT Jasaraharja Putera in light of performing certification of ISO 9001:2015

1. Introduction

Insurance currently holds a very important role in providing certainty of protection or even protection for people in terms of commercial and non-commercial use. Insurance can provide protection in the form of education, health, pension, death and wealth. One of the needs in life vital in this globalization era is the need for insurance services. This encouraged the rapid spread of insurance companies as well as causing people to worry over their life insurences. Based on this fact, a lot of new insurance companies offer various types of policy in accordance to community's needs. PT. Jasaraharja Putera took part in the

* Corresponding author: b.purwanggono@gmail.com

insurance loss field and was established on 6th of June 1996. Pt Jasaraharja is a subsidiary of PT Jasaraharja. Currently PT Jasaraharja Putera have 15 leading products including JP Graha or fire insurance. JP Graha is a fire insurance product intended for wealth protection. Whether it is a house, office, shop, or factory suffer losses or damages due to fire or other events covered by the policy. Properties eligible for JP Graha insurance included building, machine, stock, furniture, and its contents. As one of the leading products from PT Jasaraharja Putera, JP Graha is required to have a good Quality Management System (QMS) to be able to improve customer satisfaction and trust. One way to meet this demand is by applying ISO 9001 concerning quality management.

There are some differences between the ISO 9001: 2015 and previous version. The newest and most prominent difference in ISO 9001:2015 is the explisit requirements of risk-based thinking to support and improve understanding and application in the process approach that already existed on the previous version of ISO 9001. Risk is the effect of uncertainty on the expected results and the concept of risk-based thinking is always implied in the requirements article of ISO 9001:2015. This international standard makes risk-based thinking more explicit and it is realized in the requirements for establishment, implementation, maintenance and continuous improvement of the quality management system. JP Graha received Certification of ISO 9001 year 2000 version since 5th of November 2007 from SAI GLOBAL Certification Organization. Then JP Graha upgraded to ISO 9001 year 2008 risk-based thinking on 29th of September 2010. Many advantages have been felt by PT Jasaraharja Putera on the JP Graha product after applying ISO 9001. Including increased work efficiency in operational level, process and documentation is done effectively causing the work to be consistent, performance reviewed regularly and focused on target achievement, consumer trust rose and well maintained, as well as improvement of the organization's bargaining position in market competition.

PT Jasaraharja Putera suffer from a decline of consistency in applying ISO 9001:2008. This is caused by a few things, including the process of implementing ISO 9001 is quite complicated. Meanwhile, the company must prepare to upgrade to ISO 9001:2015 because ISO 9001:2008 will expire in September 2018. This causes the company difficulties in identifying readiness of the ISO 9001 upgrading process. Based on this background, the author is interested to research the readiness of JP Graha product from PT Jasaraharja Putera in upgrading ISO 9001:2008 toISO 9001:2015. This research uses gap analysis. Gap Analysis is a comparison of actual performance with potential or expected performance. In this research the author tries to compare processes done by PT Jasaraharja Putera with JP Graha in applying ISO 9001:2008 and the conditions needed in the Certification of ISO 9001:2015. So through gap analysis the author may provide the level of readiness of the company in Certification of ISO 9001:2015 and provide suggestions and recommendations to the company to increase level of readiness in the certification of ISO 9001:2015.

2. Overview

2.1 Definition of ISO 9001

ISO standard is a standard containing requirements related to quality management system published by IOS (*International Organization for Standarization*). ISO 9001 is the standard management requirement and not a standard product specification. As a standard management requirement, the contents of ISO 9001 is a series of requirement articles ensuring consistency of management process related to the quality in a system. Articles in ISO 9001 contain what must be done by the organization to implement and build each organization management system which grow into a management system elaborating how to perform a series of activities that becomes a specific requirement on each organization [1].

2.2 Quality Management System ISO 9001:2015

Renewal of ISO 9001:2015 comes with a completely different focus from the previous version that contains too many documents and forms. ISO 9001:2015 no longer question documents as much, instead it focuses on the company performance using a risk-based thinking approach and the concept of Plan - Do - Check - Action is applied at all levels of the organization [2]. This is evident by several things [3]:

1. "6 compulsory procedures" term is eliminated
2. There is no separation between the procedure and form. Both are now called documented information. This means one is represented by the other.
3. Owning a Quality Manual is now not a requirement, ISO 9001:2015 also no longer dependent on management representative considered as the only one most responsible for implementing ISO 9001. Now every part is responsible for each of their portion.

2.3 Principles in ISO 9001:2015

The quality management system of ISO 9001 : 2015 revisions have some significant changes and one of them is a change in principles of quality management from 8 to 7 the principles of quality management [4] : Customer Focus, Leadership, Involvement of people, Process Approach, Improvement, Decisions based on evidence and Relationship Management.

2.4 Gap Analysis

Gap analysis is defined by *IT Infrastructure Library* (ITIL) as an activity that compare two kinds of data and identify their differences. *Gap Analysis* is normally used to compare a set of conditions. *Gap analysis* is generally structured on a set of field, topic or category, making the *gap analysis* efficient in knowing the sector that need to be repaired. *Gap Analysis* is effective because the checklist is structured and in accordance with the subject. *Checklist* will include all existing requirements and created in the form of hierarchy in its assessments, this will include common questions and provides an overview of the topics or categories that will be assessed. Questions in the *checklist* is made in detail and make an evaluation of each individual requirement if required. Each question related to other questions to ensure searchability [5].

The following are steps in the *gap analysis*:

a. Determination of *Score*

The Score used in *gap analysis* is shown in Table 1.

b. *Checklist Assessment*

Checklist assessment by respondents based on the current condition of organization. Respondents are selected by competence. Assessmentsis based on the terms of the *scoring* described in the Table 1.

c. *Gap* Assessment

Assessment of *gap* aims to see how big of a *gap* is in the company. The percentage is obtained through the sum of *score* per variable and divided by the maximum value on the variable. The smaller the gap the better. The percentage represent company readiness in the implementation of ISO 9001:2015. Table 2 shows the range of *gap* value [6]:

Table 1. Score Gap Analysis

Score	Understanding
1	If the organization or the company does not understand what is required and does not do it.
2	If an organization or company understand the importance of the activity but does not do it.
3	If the organization or the company have the documents but have not implemented or have implemented but not recorded.
4	If an organization or company do the activity albeit inconsistently.
5	If an organization or company do the activity consistently.

Table 2. Range Gap Analysis

Percentage	Illustration
75% -100%	The organization is ready to complete the QMS ISO 9001:2015 and undertake certification.
50% - 74%	The organization must still improve the QMS for the preparation of ISO 9001:2015
1% - 49%	QMS of the organization needs much improvement due to vast differences from the Quality Management System ISO 9001:2015.

3. Methodology

3.1 Research Method

The research is done in PT Jasaraharja Putera branch in Semarang as the branch that acquired ISO 9001 on the product JP Graha, located in Jl Imam Bonjol no. 151 Semarang. The implementation of this research started on October 2017 until finished.

The method used in this research is *gap analysis*. *Gap analysis* aims to understand current situation and compare it with expected situation. In this research the point of reference for comparison is ISO 9001:2015. The current condition data is obtained by performing *internal audit checklist* ISO 9001:2015 to responsible responsible in PT Jasaraharja Putera Semarang. Including the head of the branch, Vice Head of the branch, and Human Resources Working Committee. Data on the condition is also obtained by performing direct observation to the quality system documents made by the company. The obtained data then weighted with a score of 1-5 with each percentage score presenting specified range of readiness of the company. The results then became a base to improve the readiness of the company in facing the Certification of ISO 9001:2015 and prioritized according to lowest percentage, level of ease and the scope of effect produced.

3.2 Identification of Research Variables

Research variables used in this research is taken from the clause included in ISO 9001:2015, the clause that become a variable is only clause 4 to clause 10.

4. Results and Discussion

4.1 Data Collection

Primary data is obtained through observation on field. In this research primary data is in the form of system observation and internal audit checklist ISO 9001:2015 which are made to understand the condition of company readiness for certification. There are 3 people who become respondents and they are selected for this portion because of their significant involvement in Quality Management System JP-GRAHA PT Jasaraharja Putera.

4.2 Data Processing

Data processing at this stage is to calculate the scores obtained from the *internal audit checklist*. The results of score calculation will then be used as the basis for identifying level of readiness of the company in facing the Certification of ISO 9001:2015. The total score for each variable from the results of the *checklist* are Organization Context (80%), Leadership (84%), Planning (75%), Support (81%), Operations (80%), Performance Evaluation (85%), and Improvement (83%).

4.2.1 Checklist Result Analysis of the Organization Context Clause

After gap analysis is performed in the company for the organization context, a value of 80% is obtained. Organization Context Clause Recap are 4.1 Understand the organization and its context (80%), 4.2 Determine the scope of quality management system (77%), 4.3 Quality management system and its processes (90%), 4.4 Quality management system (QMS) and its processes (73%) and Readiness Value (80%).

4.2.2 Checklist Result Analysis of the Leadership Clause

After gap analysis is performed in the company for the leadership, the value of readiness obtained is 84% The value on each point has exceeded 75 percent which indicates that the company is ready to do certification in terms of leadership. Table 5 below shows the results of gap analysis performed on the leadership clause. Leadership Clause Recap are 5.1 Leadership and commitment (86%), 5.2 Policies (85%), 5.3 Role of the organization, responsibility and authority (80%) and Readiness Value 84%.

4.2.3 Checklist Result Analysis of the Planning Clause

There are 3 sub-clauses in clause 6.1 planning. Namely : 6.1 Actions to deal with risks and opportunities, 6.2 Quality objectives and plans to achieve it, and 6.3 Planning changes. There is 1 sub-clause that is under the standard value, namely sub-clause 6.1 Actions to deal with risks and opportunities. So the company needs to pay attention and perform improvement on this sub-clause. Planning Clause Recap are 6.1 Actions to deal with risks and opportunities (60%), 6.2 Quality objectives and plans to achieve it (86%), 6.3 Planning Changes (80%) and Readiness Value (75%).

4.2.4 Checklist Result Analysis of the Support Clause

There is no sub-clause with a value under 75%. The following is an analysis for each sub-clause on clause 7. Support Clause Recap are 7.1 Resources (79%), 7.2 Competencies

(85%), 7.3 Awareness (80%), 7.4 Communication (80%), 7.5 Information documented (82%) and 81%.

4.2.5 Checklist Result Analysis of the Operational Clause

The company has performed activities on this clause good enough, but it still needs some improvement because there are some sub-clauses with value below standard. The following is an analysis for each sub-clause on Operational clause. Operational Clause Recap are 8.1 Planning and control of operations (90%), 8.2 Requirements for the products and services (85%), 8.3 Design and development of products and services (69%), 8.4 External products and services control (8.5), Production and service provision (84%), 8.6 Release of product and service (70%), 8.7 Control over output that does not match (83%) and Readiness Value (80%).

4.2.6 Checklist Result Analysis of the Evaluation Clause

There are no sub-clause below the standard value. This shows that the implementation of this clause is good enough. The following is an analysis for each sub-clause on Evaluation clause. Evaluation Clause Recap are 9.1 Monitoring, measurement analysis and evaluation (85%), 9.2 Internal Audit (88%), 9.3 Management Overview (82%) and Readiness Value (85%).

4.2.7 Checklist Result Analysis of the Improvement Clause

There is no sub-clause with value under 75% on this clause. The following is an analysis for each sub-clause on Improvement clause. Improvement Clause Recap are 10.1 General (88%), 10.2 Incompatibility and repair actions (78%) and Readiness Value (83%)

4.3 Recommendations

Based on the results on section 4.2, it is known there is some clause that has low value of readiness in light of preparation for the Certification of ISO 9001:2015 which require improvements. The following is recommendations concerning the clauses with low readiness value:

Table 3. Recommendations

Clause	Improvement
4.2 Determine the scope of the quality management system (QMS)	The company must clearly define the boundaries of what is inside the scope of QMS Create documents related to the scope of QMS because this document is also prerequisite of ISO 9001:2015
4.3 QMS and its processes	The company needs to clearly define the risk and opportunities of every business process
5.3 Roles of companies, responsibility and authority	Top management of the company especially Quality Management Repressentative (QMR) start to Adjust and ensure all processes on the Quality Management System are in accordance with the needs of ISO 9001:2015 and perform repairs or adjustments if there are incompatibilities. QMR should also form a small team in preparation for the Certification of ISO 9001:2015
6.1 Actions to deal	Risk and opportunities must be clearly defined for all

with risks and opportunities	process Add one division on the organizational structure of the company which is the risk management division The company will need to implement and integrate risk management into quality management system and evaluate its effectiveness The company should create documents related to risks and opportunities although is not prerequisite to ISO 9001:2015. But this document may become a source of information about what is considered and what has been done. This document will also facilitate the audit process
7.1.6 company knowledge	Continue to update company knowledge, in terms of written knowledge such as standards, results of their discussions with customers, discussions with academics, the company should archive this information to make it easily accessible. For internal knowledge can inferred from an expert who works at the company, experience from failures and others
7.5 Documented Information	Matching documents and records currently used by the company to be the documented information in accordance with the needs of ISO 9001:2015.
8.3.3 Input design and development	The company will need to make documented information in the form of input recording of the design and development because it is a prerequisite of ISO 9001:2015
8.3.4 Control design and development	The company will need to make documented information in the form of the activity of the control of design and development recording
8.3.6 Change in design and development	Improve the consistency of the company in the reviewing and identifying changes during or after the development is done. The company will need to make documented information in the form of changes in design and development recording
8.4.2 Types and control level	Ensure that all external parties who have impact on the company is in the scope of control of the Quality Management System (QMS) by performing periodic evaluation on external parties.
8.6 Release of product and services	The company will need to make the documented information in the form of compliance recording
9.2 Internal Audit	Changing the company internal audit sheet with the ISO 9001:2015 internal audit sheet Maintain and improve the consistency of the company in conducting periodic internal audit
10.2Incompatibility and repair actions	Immediately change QMS ISO 9001:2008 into QMS ISO 9001:2015 because ISO 9001:2008 will *expire* on November 2018

5. Conclusions

https://doi.org/10.1051/shsconf/20184901005

Based on the results of thorough gap analysis on the product JP-GRAHA PT Jasaraharja Putera concerning readiness for certification of ISO 9001:2015, the company has value readiness of 80%. This shows that the company is of category 1 (75-100%) namely the company or organization is ready to do QMS ISO 9001:2015 and undertake certification.

There are several improvements that can be done by JP-GRAHA PT Jasaraharja Putera to improve readiness in facing certification of ISO 9001:2015. The improvement can be seen in details on table 13 to table 18.

References

[1] S. L. Prabowo, "Implementasi Sistem Manajemen Mutu (ISO:9001:2008) di Perguruan Tinggi," *Guideliness IW A-2,* 2009.

[2] V. Gasperz, "Total Quality Management," *PT. Gramedia Pustaka Utama,* 2001.

[3] L. Ibrohim, "ISO 9001 2015 FDIS Penjelasan Klausul- Klausul.," *PT. Gunastara,* 2015.

[4] A. Parakarsa, N. Setyanto dan L. kusuma, "Analisis Penerapan Sistem Manajemen ISO 9001:2008 Menggunakan Gap Analysis Tool (Studi kasus PT. Sahabat Rubber Industries, Malang)," *Jurnal Rekayasa dan Manajemen Sistem Industri,* 2015.

[5] M. Picard, A. Renault, B. Barafort dan S. Cortina, "Measuring readiness for compliance: A gap analysis tool to complete the TIPA process assessment framework.," *Springer international publishing Switzerland,* Vol. %1 dari %2633(2016),106-116, 2016.

[6] A. Bakhtiar dan B. Purwanggono, "Analisis Implementasi Sistem Manajemen Kualitas ISO 9001 : 2000 dengan menggunakan GAP Analysis Tools (Studi Kasus PT. PLN (PERSERO) PILITRING JBN Bidang Perencanaan)," *Jurnal Teknik Industri Undip,* 2009.

Food Safety System Design Using Hazard Analysis Critical Control Point (HACCP) on Beverage Product X at a Milk Tea Producer

Praditya Ajidarma[1,*], *Muthya* Islamiaty[1], *Fariz* Hasby[1], and *Dradjad* Irianto[1]

[1]Manufacturing System Research Group, Bandung Institute of Technology, Jalan Ganeca No. 10 Bandung, West Java, Indonesia

Abstract. Food safety is a primary concern in food industry. Hazard Analysis Critical Control Point (HACCP) is a systematic preventive approach that aims to ensure food safety in a comprehensive manner, starting from raw material retrieval from supplier, production phase, up to the distribution to end consumers. Furthermore, the approach is capable of reducing risks associated with food hazard to their acceptable level. Our research took place in a company that produces a wide array of milk tea with a variety of flavours, i.e. product X. Recent increase in overall demand and competitors in Indonesian milk tea market subsequently forces the company to pay more attention to the quality of their product X. Thus, the company aspires to certify its food safety system and management. This research proposes a design and implementation plan of HACCP for quality assurance and food safety of product X. The main outcomes of this research are the HACCP plan for raw material, packaging material, and production process. Further potential improvements are suggested with regard to production process, production equipment, production layout, human resource, sanitation, and documentation system. The HACCP plan consists of ten Critical Control Points (CCPs), a series of documentation forms, and additional documents specifically attributed to support the implementation phase of HACCP plan.
Keywords: food safety, quality management, HACCP, hazard analysis

1 Introduction

Standards are conceptually introduced as an enforcement tool for quality in different majors in higher level of education [1]. Furthermore, sustainability and environment have been key aspects in recent development of engineering and design [2-3]. This research aims to complement student's conceptual understanding by providing a case study of the implementation of standards in food industry that is profoundly concerned with sustainability and environmental issues.

Food safety management system has been proven to result in higher customer's satisfaction, better food safety, and increased perceived quality in company's branding [4].

* Corresponding author: ajidarmap@itb.ac.id

One implementation of standards in food safety management is HACCP. Theoretically, HACCP is a part of ISO22000, and most companies are advised to acquire a HACCP certification first-hand prior applying for ISO22000 [5].

This research takes place in a certain milk tea producer, the producer of Beverage Product X that is based in Bandung. Due to the perishable nature of its products, the company aims to standardize its food safety by acquiring ISO22000 certification, starting from the HACCP requirements.

2 ISO14001 Environmental Management System

HACCP is a preventive system that identifies, measures, and controls the hazard level in a food system, with an objective of satisfying food safety requirements in a company. In practice, HACCP complements the implementation of ISO9000, quality management system, and even serves as the basis of ISO22000, food safety management certification [6]. There are 4 key steps of preparing and applying HACCP in a company [7]:

1. Planning, which includes a gap analysis
2. Doing, in which the company applies the 12 implementation phases of HACCP
3. Checking, or HACCP plan verification prior full implementation
4. Acting, when HACCP plan, prerequisite program, and other food safety management programs have been implemented and monitored

There are 12 steps of HACCP application according to SNI 01-4852-1998. Out of those steps, five are of preparation phase and the remaining seven are general hygiene principles. The 12 steps of HACCP as [8] is listed in Table 1.

Table 1. Twelve Steps of HACCP (Thaheer, 2005)

HACCP Preparation		**HACCP Principle**	
Step	Program	Step	Program
1	Assemble the HACCP Team	6	Conduct a Hazard Analysis
2	Describe the Product	7	Determine Critical Control Points
3	Identify the Intended Usage	8	Establish Critical Limits for Each CCP
4	Construct Flow Diagram	9	Establish Monitoring System for CCP
5	On-Site Confirmation of the Flow Diagram	10	Establish Corrective Actions
		11	Establish Verification Procedures
		12	Establish Documentation and Record Keeping Procedures

https://doi.org/10.1051/shsconf/20184901006

3 Methodology

The methodology of this research is in accordance with HACCP principles adopted from Codex: Alimentarius Commission – WHO, which consists of 7 general hygiene principles, Codex practical principles, and other relevant food safety regulations. First, preliminary study is conducted on both theoretical methods and practical implementation of the case. Furthermore, the five steps of HACCP preparation are executed, which end with a confirmed flow diagram. The methodology proceeds with the application of HACCP principles, starting from hazard analysis until the design of record-keeping procedures. Lastly, recommendations from gap analysis and a detailed implementation plan are proposed.

4 Analysis

Prior analysis, required data is gathered based on the twelve HACCP steps. The gathered data consists of consumer identification data and production flow diagram of the products. Further, the flow diagram is validated directly with on-site observation on production floor. An example of confirmed production flow diagram for mixing sub-process is presented on Fig. 1.

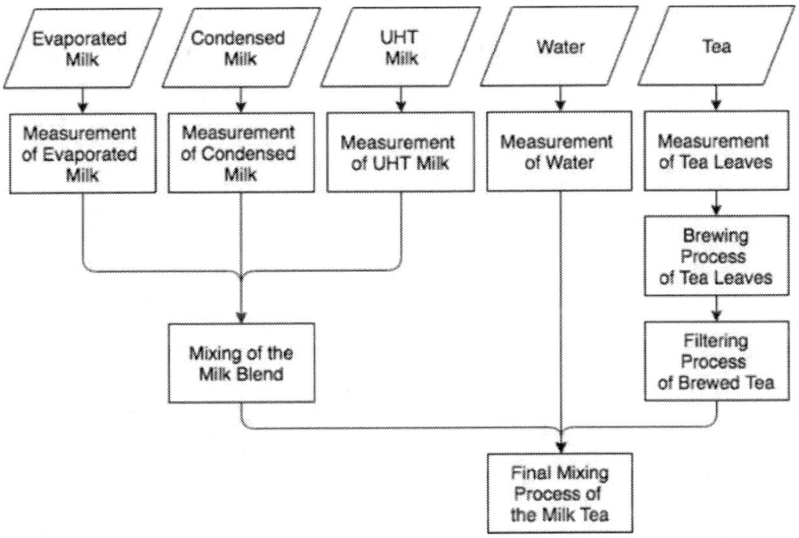

Fig. 1. Process Flow Diagram for the Mixing Sub-process of Product X

After the data is gathered, gap analysis is conducted. In this event, the company's current business model is compared to the referred standards, i.e. ISO 22000, and thus the gap between the two conditions is analyzed. This information is crucial for the following stages, which are HACCP plan design and the fulfillment analysis.

4.1 HACCP Safety Management System Design

The design of food safety management system begins by implementing the HACCP principles, or the sixth HACCP step. After the entire process flow diagram is verified on-site, this research proceeds to hazard analysis and identification. The hazard identification refers to several standards and food regulations, three of which are KBPOM No.

HK.00.06.1.52.4011, SNI 7387:2009, and SNI 3141.1:2011. Further, the significance of each potential hazard is calculated based on its likelihood and severity, in accordance with ZNS standard of ISO 31000:2009.

The critical control points (CCP) are determined based on the process flow diagram and hazard analysis. In this research, 4 CCPs are identified in raw material handling and 6 CCPs are identified within the production process. Moreover, critical limit and monitoring system are designed for each CCP, accordingly. To minimize potential hazards, the monitoring system defines three specific components for each point: procedure, frequency, and person in charge. Afterwards, the entire plan is verified using sampling, audits, and document verification.

Lastly, documents and record-keeping procedure is defined properly. Appropriate documentation benefits the company when it requires a data-driven root cause analysis, CCP monitoring, or performance appraisal for certification.

4.2 Fulfilment Analysis of Prerequisite Programs

The purpose of prerequisite programs is to guarantee the continuity of the production process and wellness of the environment during the implementation of HACCP. This research analysed the fulfilment of two prerequisite programs: Good Manufacturing Practices (GMP) Analysis and Sanitation Standard Operation Procedures (SSOP).

The GMP is a system that ensure each product to be produced and controlled according to its required standards. The GMP analysis is based on 75/M-IND/PER/7/2010 food processing regulation of Indonesian Ministry of Industry [9]. Based on such standard, there are six aspects that are currently unfulfilled: construction, laboratory, product labelling, sanitation and maintenance, documentation, and human resource training.

Furthermore, the analysis on SSOP is conducted to control, monitor, and improve the sanitation aspect of the factory and its surrounding environments. The SSOP analysis is based on 715/MENKES/SK/V/2003 food hygiene and sanitation regulation of Indonesian Ministry of Health [10]. This research also considers regulations released by U.S. Food and Drug Administration (FDA) to complement the SSOP analysis. Our analysis indicates that two aspects are dissatisfying: water safety and pest control.

4.3 Managerial Implications

Based on the analysis on HACCP implementation, prerequisite programs fulfillment, and the existing condition of the company, this research recommends potential improvements on different aspects of the company. The recommendations are made on the following aspects: production procedure, layout, human resource, sanitation, and record-keeping system, as displayed on Table 2. Additionally, implication analysis is conducted on both managerial and business aspects of the plan.

Table 2. Proposed Improvements on Four Aspects

Aspects	Recommendations
Production Procedure	Raw material quality checklist, minimum distance of cooling storages, additional warning sign on product label, and assignment of food safety supervisor
Layout and Construction	Eliminating dead corner to reduce bacterial growth

Aspects	Recommendations
Human Resource	HACCP implementation training and HACCP-specific internal audit
Sanitation	Instalment of anti-siphon valves and procurement of pest control equipment

5 Conclusion

In this research, a HACCP-based safety management system is designed in a milk tea producer. Further, there are three key outcomes that this research concludes: the HACCP plan as the main outcome of this research, a newly-designed documentation system to reinforce implementation, and recommendation points to ensure effective implementation of HACCP.

First, the output of the HACCP-based safety management system is a HACCP plan. The objective of the plan is to maintain the quality and food safety of product X. The plan consists of ten critical control points, each with its own critical limit, monitoring requirements, corrective actions, verification, and record-keeping procedures.

Secondly, this research also proposes a documentation system that aims to control the effectiveness of HACCP implementation. Further, it also enables user to track food hazards and to keep the implementation records in general. The newly-designed documentation system consists of 30 documents, which includes 1 quality manual, 13 procedures, 9 work instructions, and 7 forms.

Lastly, based on the current situation of the company, several improvements are recommended to ensure an effective implementation of HACCP. Thus, this research provides a set of specific recommendations for CV Bumi Lestari, which aims to improve its production procedures, production equipment, production layout, human resource, sanitation, and documentation system.

References

1. D. Irianto, A. Maruf, and T.M.A.A. Samadhi, A Quest for Curriculum of Standardization Education Program in Engineering Higher Education in Indonesia, *ICES Conference & WSC Academic Day* 2012, Bali, ISSN: 2252-9357. (2012)
2. D. Irianto, P. Paramitha, Implementing Design for Six Sigma in Green Manufacturing: A Case at Food Industry, *Proceedings 10th International Conference on Quality in Research*, ISSN: 1411-1284. (2013)
3. D. Irianto, Sustainable Manufacturing, Keynote Speech, *International Conference on Industrial Engineering: Theory Methodology and Application*. (2015)
4. P. Sampaio, & S. Teixeira, Food Safety Management System Implementation and Certification: Survey Results, *Total Quality Management & Business Excellence*, 24:3-4, 275-293 (2012)
5. T. Wahono, Pelatihan HACCP. Malang: Universitas Brawijaya. (2002)
6. Codex Alimentarius Commission, Hazard Analysis and Critical Control Point (HACCP) System and Guidelines for Its Application (1997), retrieved from http://www.fao.org/docrep/005/y1579e/y1579e03.htm. on 1 January 2018
7. S. Mortimore, C. Wallace, HACCP: A Practical Approach (3rd edition), New York: Springer (2013)

8. H. Thaheer, Sistem Manajemen HACCP, Jakarta: PT Bumi Aksara (2005)
9. Indonesian Ministry of Industry, Peraturan Menteri Perindustrian Indonesia tentang Pedoman Cara Produksi Olahan yang Baik Nomor 75/M-IND/PER/7/2010, Jakarta: Indonesian Ministry of Industry (2010)
10. Indonesian Ministry of Health, Keputusan Menteri Kesehatan Republik Indonesia tentang Persyaratan Hygiene Sanitasi Jasaboga Menteri Kesehatan Republik Indonesia Nomor 715/MENKES/SK/V/2003, Jakarta: Indonesian Ministry of Health, (2003)

Analysis of implementation of ISO 9001:2015 in SME Aldin Natural Home

Naniek Utami Handayani[1,*], *Anita* Mustikasari[2,*], *Sabrina* Hutagalung[1] and *Baby* Monica[1]

[1]Department of Industrial Engineering, Diponegoro University, Semarang, Indonesia 50275
[2]Department of Management, Yogyakarta State University, Yogyakarta, Indonesia 55281

Abstract. SME is required to improve their service quality and resource management for giving customer loyalty toward product offered. The ISO 9001 is one of quality management system standard which is approved globally which are frequently communicated thru service quality insurance to solve the matter of indistinct expectancy Aldin Natural Home is SME engaged in dolls production. In the implementation of quality management system, they have lack on standard operating procedure (SOP). This study aims to analyze and find out the readiness of ISO 9001:2015 implementations in SME Aldin Natural Home. Gap analysis method also is conducted in this research to describe the differences between required and actual quality management systems. The results are SME Aldin Natural Home need to understand and review of implementation of ISO 9001:2015 with the highest 29% on clause 8, Operation while the lowest is 14% on clause 9, Performance Evaluation.

1 Introduction

In the globalization era, SME is required to improve their service quality and resource management for giving customer loyalty toward product offered. Therefore, product guarantee strategy is needed to gain customer trust.

The ISO 9001 is one of quality management system standard which is approved globally which are frequently communicated thru service quality insurance to solve the matter of indistinct expectancy [1, 2]. More than one million organizations in 187 countries has implemented ISO 9001 standard, and a high number of these organizations purpose at attaining product innovation [3]. According to Gaspersz 2001, the ISO 9001 set up requirement and recommendation for design and assessment of quality management system that aims to ensure companies or SMEs give products (goods/ services) which meet the specified requirements [4]. ISO 9001 objective to increase customer satisfaction by means of the effective use of the system, including processes for continued improvement

A recent version of International Organization for Standardization (ISO) 9001 raise on September 2015, or it known as ISO 9001:2015. For the first time, requirements are released that have contents a responsibility towards organizations to ponder the organizational knowledge role as a resource. This details will oblige organization to consider all the phase of the knowledge management cycle systematically [5].

The differences between ISO 9001:2008 and ISO 9001:2015 located on risk management. In ISO 9001:2008, risk is considered exclude in system whilst in ISO 9001:2015, considered include in system.

Aldin Natural Home is SME engaged in dolls production. In the implementation of quality management system, they have lack on standard operating procedure (SOP). For instance, using old version of SOP however there are modification or addition products. Therefore, it is needed more detailed production record and further review to eliminate the lack between real work and SOP. In addition, it will support SME Aldin Natural Home to receive ISO 9001 certification.

This study aims to analyze and find out the readiness of ISO 9001:2015 implementations in SME Aldin Natural Home. Gap analysis method also be conducted in this research to describe the differences between required and actual quality management systems.

2 Research Method

Gap analysis is method that could identify discrepancy between ISO 9001 standard and SME organization process. Therefore, with that method, organization could take actions or strategies to eliminate lack that occur.

Gap analysis checklist is one of ways that simplify to analyze factors related ISO 9001 standard in form of checklist. Checklist is arranged by ISO 9001:2015 standard. In order to simplify analysis each clause so it is given assessment indicators in each clause as follow or shown on Figure 1:

Score 1 : The organization or SME does not understand what is required and does not do so.

Score 2 : The organization or SME understands the importance of the activity but does not do so.

Score 3 : The organization or SME has documents but has not been applied or done but not recorded.

Score 4 : The organization or SME is engaged in an activity but not consistent.

Score 5 : The organization or SME is doing well (done consistently).

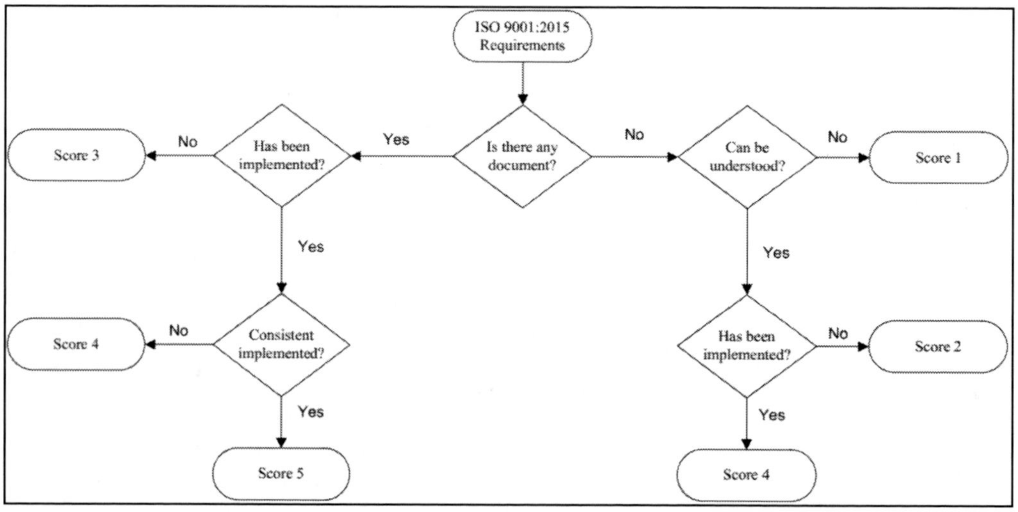

Fig. 1. Assessment Indicators

3 Results and Analysis

3.1 Analysis of Implementation ISO 9001:2015

The ISO 9001: 2008 consists of 8 clauses, scope, normative references, terms and definitions, quality management system, management responsibility, resource management, product realization, measurement, analysis and improvement [6]. While the ISO 9001:2015 consists of 10 clauses, scope, normative references, terms and definitions, context of the organization, leadership, planning, support, operation, performance evaluation and improvement [7].

SME Aldin Natural Home has not been implemented ISO 9001:2015. So, this research assesses readiness of SME Aldin Natural Home in implementation of ISO 9001:2015 using questionnaire as shown on Table 1.

Table 1. Questionnaire Result Related on ISO 9001:2015 Clauses SME Aldin Natural Home

No Clause	Clause	Score	Maximum Score	Percentage
4	Context of the Organization	5	20	25%
5	Leadership	4	15	27%
6	Planning	3	15	20%
7	Support	7	25	28%
8	Operation	10	35	29%
9	Performance Evaluation	2	15	14%
10	Improvement	3	15	20%

From the assessment results can be seen the implementation of ISO 9001: 2015 with range as shown Table 2. It shows that SME Aldin Natural Home is in range 14% to 29% so it needs to understand and review of implementation of ISO 9001:2015 because there are many lacks of standard.

Table 2. Range Assessment Result

Percentage	Details
100%	Ready for implementing ISO 9001:2015
91%-99%	Ready for implementing ISO 9001:2015 but there are several work procedures inconsistent
76%-90%	Ready for implementing ISO 9001:2015 but there are several work processes has not proceed
51%-75%	Many work procedures has not executed but has been implemented ISO 9001:2015
0%-50%	Company or SME needs to understand and review of implementation of ISO 9001:2015 because there are many lacks of standard.

3.2 Analysis of Obstacles of ISO 9001:2015 Implementation

Several obstacles faced by SME Aldin Natural Home in implementing of ISO 9001:2015 shown on Table 3.

Table 3. Obstacles Related on ISO 9001:2015 Clauses SME Aldin Natural Home

No Clause	Sub-Clause	Obstacles
4	4.1 Understanding context of organization 4.4 Quality management system and their processes	1. Lack of understanding to define issues related with SME. 2. Has not had criteria and method that needed to ensure effective operation.
5	5.1 Leadership and commitment 5.2 Policy	1. Lack of commitment to do controlling toward management system 2. Lack of job desk because of less communication among top management.
6	6.1 Action to overcome risk and opportunity 6.2 Quality objective and planning to reach it	1. There is no planning to resolve risk, even SME has less attention toward the bad impact that will occur on production process 2. Lack of consistency on reach their target
7	7.1 Resources 7.2 Documented information	1. It has limited time because only has few competent employees 2. Limited facilities
8	8.2. Requirements for products and services 8.3 Design and Product Improvements 8.4 Product and Services External Control	1. Limited of resources, there is no work standard regarding on product and job desk itself. 2. This SME use make to order method, it could not handle big quantity order so lack of time to fulfill the order 3. There is no innovation and improvement toward their product 4. Less of promotion
9	9.1 Monitoring, measurement, analysis and evaluation 9.2 Internal Auditing	1. There is no specific work that has to monitor. 2. Less of work control, there is no certain procedure to make products 3. Do not use personal protective equipment in making products
10	10.2 Incompatible and corrective action 10.3 Continuous improvement	1. There is no exact procedure so it is not known whether the work process is in accordance with the procedure 2. It has no desire to do continuous improvement

3.3 Improvement Suggestion

SME Aldin Natural Home needs commitment and effort from management in order to it could improve quality management system to get ISO 9001:2015 certifications. For clause 4, context of organization, SME Aldin Natural Home should give training for top management to improve their quality in analysis related issues to minimize repeated problems. Also, for employee, they give knowledge about how to important increasing quality, how strategy to reach it. Improvement suggestions for clause 5 are doing management observation and giving relevant assignment. With those recommendations, top management has an authority and responsibility and ensures each process done according to procedure. While, for clause 6, SME Aldin Natural Home need give training about risk management and review company goals to identify risk and take an action to prevent risk. Clause 7, about support, the SME should improve internal communication, make clear document and also add resources. It would overcome problem in limited time and knowledge and support for auditing. Increasing promotion is becoming one of suggestion that recommended for solving clause 8. Also, to overcome problem in clause 9, performance evaluation, the SME Aldin Natural Home need making assessment form and quality internal audit to ease evaluation process and further improvement and to assess whether product quality is increasing or decreasing. Last, for clause 10, analysis of profit increment and defect analysis are needed to measure whether work procedure is efficient or not. In addition, from those improvement suggestions, the efficient procedure can be assessed through quantity of defect.

4 Conclusions

Based on assessment results, SME Aldin Natural Home is in range 14% to 29%. The details are as follows context of the organization 25%, leadership 27%, planning 20%, support 28%, operation 29%, performance evaluation 14% and improvement 20%. Therefore, it needs to understand and review of implementation of ISO 9001:2015 because there are many lacks of standard.

There are several obstacles that faced in each clause ISO 9001:2015. In clause 4, lack of understanding and there are no readiness to implement ISO 9001:2015. Clause 5 has difficulty on lack of commitment and lack of division of work. Next, clause 6, there is no planning to prevent risk. In clause 7, it is limited time and facilities. Then, clause 8, it is limited on resources and less of promotion. Clause 9 has problems on less controlling in production process and last, clause 10, it found less of knowledge and no desire to do continuous improvement.

In order to overcome the obstacles, there is several improvement suggestions related on ISO 9001:2015 clauses SME Aldin Natural Home. Clause 4, it is giving training for top management and employees. Clause 5, do reviewing management and making suitable task. In clause 6, giving risk management training. Clause 7, there is improving internal communication, making clear document and adding resources. In clause 8, increasing promotion and making clear procedure. Next, clause 9, there is making assessment form and making quantity internal audit. Last clause 10, do profit increment and defect analysis.

References

1. H. P. Yaya, L., F. Marimon, and M. Casadesus, Ind. Manage. Data Syst. **111**, 1194 (2011)
2. H. P. Yaya, L., F. Marimon, and M. Casadesus, Ind. Manage. Data Syst. **113**, 1206 (2013)
3. B. Manders, H.J. de Vries, and K. Blind, Technovation **48,** 41 (2016)
4. V. Gaspers, Gramedia Pustaka Utama (2001)
5. J.P. Wilson and L. Campbell, J. Knowl. Manag. **20**, 829 (2016)
6. H. Karahalios, Safety Sci. **63**, 104 (2014)
7. D. Hoyle, Rout. (2017)

Effectiveness analysis of ISO 9001:2015 implementation at manufacturing industry

Ahmudi[1*], *Bambang* Purwanggono[1], *Naniek* Utami Handayani[1]

[1]Industrial Engineering Department, Faculty of Engineering, Diponegoro University, Jl. Prof. Soedarto, Tembalang, Semarang, Indonesia

Abstract - Globalization era requires many companies to improve service quality and resource management capabilities in order to gennerate public confidence in the products and commitments offered. The quality improvement program is contained in Quality Management System (QMS). QMS is developed in ISO 9001 standart certification.ISO 9001 is a standard that contains quality management system requirements issued by the International Organization for standarization (IOS). Companies that become the obyect of research is PT Sejahtera, this company is engaged in manufacturing.the company suffered several losses in the form of withdrawal of order from customer. Besides that happened also complaint from customer will product which yield not according to standard which have been approved. This research aims to evaluate the effectiveness of ISO 9001: 2015 implementation at PT. Sejahtera. The method used is Gap Analysis. Calculations on gap analysis use checklists made based on requirements of ISO 9001: 2015.the result of calculation using checklist will show how big the effectiveness of PT. Sejahtera in implementing ISO 9001: 2015

Keywords: Industry,Gap Analysis, ISO 9001: 2015, Quality Management System.

1. Introduction

The globalization era requires many companies to improve service quality and resource management capabilities in order to generate public confidence in the products and commitments offerred [1]. To obtain public trust is required strategy of product quality assurance. Quality management can be run through Total Quality Management (TQM) because according to Feigenbaum [2] quality is an important joint that determines the success or failure of a business in today's market-oriented quality achievement, the quality itself is not the field of key business strategy and an important factor in business strategy planning.

To improve efficiency, competitiveness and customer satisfaction, many companies are adopting Quality Management System (QMS). QMS is widely developed in ISO 9001 standard certification. The purpose of ISO 9001 is to help companies to implement and run

* Corresponding author: ahmudi.80@gmail.com

an effective QMS by enhancing the company's ability to design, produce and deliver quality products and services. Based on Fonseca (2015) there are already more than one million companies in the word that have achieved ISO 9001 standard certification. It shows that ISO 9001 certificate gives advantages to internal company, in the form of product quality and good company performance [3]. Jan and Lin on Psosmas [3] implementing ISO 9001, have a direct positive impact on operational performance. From a customer perspective, ISO 9001 implementation creates a positive sentiment from customers toward the manufacturing industry, and encourages customers to provide long-term support, and increase customer satisfaction and loyalty [3].

This research was done at PT Sejahtera which was established in 2011 is a company engaged in manufacturing false eyelashes, 100% of its products are exported abroad including America, Europe, Asia and Afrika. Although it has been implementing ISO 9001 since 2013, there are still complaints from customers regarding the quality of the products. The products are not in accordance with the approved standards. This study aims to analysis and evaluate the effectiveness of ISO: 2015 implementation by using gap analysis. The output of this research is to input the evaluation of Quality Management System (QMS) to top management so that the decision will be taken on target so that complaints from customers can be done well and quickly.

2. Literature Review

2.1 Definition of ISO 9001

ISO 9001 standard is a standard that contains requirement related to the quality management system issued by IOS (International Organization for standarization). ISO 9001 standard is a standard management requirement and is not a product specification standard. As a standard of management requirements. The contents of ISO 9001 is a series of requirements clauses that ensure consistency of management processes related to quality in a system. The article on ISO 9001 contain what should be done for organizations that want to implement and build a management system that explains how to conduct a series of activities that are specific to each organization [4].

2.2 Quality Management System ISO 9001: 2015

ISO 9001: 2015 is the result of a five-year revision of ISO 9001. The revision process of ISO 9001: 2015 was started by ISO/Technical committe (ISO/TC 176) aimed at ensuring international standards. ISO 9001: 2015 reflects changes in an increasingly complex and dynamic environment. ISO 9001: 2015 consistens of ten clauses, namely the two main parts and the attachment located in clause zero to clause three and the part of the conditions contained in clauses four to clause ten. This clause is designed according to the structure in annex SL, which is a High Level Structure (HSL) which is the same basic reference for all management system structure published by IOS [5]. The clauses of ISO 9001: 2015 are: scope, normative references, term and definitions, organizational context, leadership, planning, support, operational, job evaluation and improvement.

2.3 Risk Based Thinking

What's new with ISO 9001: 2015 is the explicit requirement of risk-based thinking to support and improve understanding and application in process approaches that already exist in previous versions of ISO 9001 standard [5]. Risk management is a process to identify the

https://doi.org/10.1051/shsconf/20184901008

hazards that exist in the business as well as the choice of ways to determine the methods and techniques used to protect the business from such hazards. The process of risk management is the identification of risk, measurement, risk control and risk mitigation [4].

3. Research Method

The research method used in this research is descriptive method. This method is used to describe something that is happening at the present moment. Descriptive method is also a study to make improvements to the current conditions that are deemed necessary to be improved.

3.1 Gap Analysis

Gap analysis has a sense of the current state and the desires or conditions of purpose and the cause of the gap between current condition and purpose conditions [6]. By doing a gap analysis, we can identify what we need to bridge the existing gap [7]. The first step of this tool is to develop a gap analysis checklist that serves to identify gaps between written procedures and processes performed. This checklist is based on the requirements of ISO 9001: 2015. To facilitate the analysis of each clause it is given an assessment indicator on each clause. Once clearly known about the gap (gap) in the company, then steps can be taken to reduce the gap.

a. Score 1: if the company does not understand what is needed and does not do so.
b. Score 2: if the company understands the importance of the activity but does not do it.
c. Score 3: if the company has documents but has not been applied or done but not recorded.
d. Score 4: if the company does the activity but is not consistent
e. Score 5: if the company is doing well (done consistently)

3.1.1 Checklist Assessment

Checklist Assessment by respondents based on current organizational conditions. The respondents chosen were the respondents who had enough competence. Assessments made under the scoring conditions described above.

3.1.2 Gap Assessment

Gap assessment aims to see how big the gap is in the company. The percentage value is obtained by summing the score per variable and dividing it by the maximum value in that variable. The smaller the gap is, the better. Table 1 shows the percentage range of gap value.

3.2 Research Steps

To get solution of problem which under study required steps that must be passed so that research can be done systematically. The steps undertaken in the research are as follows:

1. Preliminary studies
   PT Sejahtera got complaint order from customer because standard order was not in accordance with which have been approved. This has an impact on recall order from customer.
2. Research purposes

Evaluate the effectiveness of ISO 9001: 2015 implementation and provide recommendations for improvement of ISO 9001: 2015.

3. Literature review
Quality Management system (QMS), ISO 9001: 2015 and Gap Analysis
4. Collect data
Interview and observation
5. Result and Analysis effectivity of ISO 9001: 2015 implementation
6. Conclusion

Table 1. Percentage range of gap value

Presentation	Descriptions
100%	Work procedures and requirements are well executed
91% - 99%	Work procedures and requirements are run but not consistent
76% - 90%	Some work procedures have not been executed but other requirements are well executed
51% - 75%	Many work procedures have not been executed but some requirements that do not require working procedures have been applied
0% - 50%	Very poor implementation. Companies need to understand and review the implementation of QM ISO: 2015 because it is still far from the requirements

4. Result and Analysis

4.1 Collect Data

Primary data obtained by conducting interviews and quesionnaires whose contents are checklist based on ISO 9001: 2015 requirements. The questionnaire is filled by each department head that exist in the company. The department involved are HRD, marketing, purchasing, inventory, production, PPIC, IT, QC and QA.

4.2 Data Processing

The result of gap analysis implementation of ISO 9001: 2015 at PT Sejahtera using checklist based on ISO 9001: 2015 requirement. Evaluation is done based on the results of questionnaires filled by each department head of company. The clauses assessed are clause 4 to clause 10. The assessment results are shown in the Table 2 and Table 3.

5. Result and Analysis

5.1 Analysis of Organization Contect Clause

Clause organizational contect get value 87%. The means that some work procedures have not been executed but other requirements are well executed. The company's constraint to clause 4 is the lack control from management representative regarding quality management

 https://doi.org/10.1051/shsconf/20184901008

system implementation. Table 4 is the result of gap analysis performed of organization contect clause.

Table 2. Gap Analysis Result

Clausul	Production	PPIC	QA	QC	Purchasing	Inventory	Marketing	HRD	IT	Total	Maks Score
4. Organization Contect	39	39	39	38	39	40	38	42	38	352	45
5. Leadership	38	41	35	35	38	38	38	41	32	336	45
6. Planning	33	34	34	34	31	34	33	36	33	302	40
7. Support	38	42	39	41	42	41	39	46	40	368	50
8. Operation	34	32	31	33	32	35	32	33	34	296	35
9. Performance Evaluation	31	33	33	32	33	34	32	35	30	293	35
10. Improvement	26	28	24	27	23	27	27	28	27	237	30

(Source: Data Calculations Result of Researcher)

Table 3. Percentage Gap Analysis Result

Clausul	Max Score (X 9)	Result Score	%
4. Organization Contect	405	352	87%
5. Leadership	405	336	83%
6. Planning	360	302	84%
7. Support	450	368	82%
8. Operation	315	296	94%
9. Performance Evaluation	315	293	93%
10. Improvement	270	237	88%
		Mean	**87%**

(Source: Data Calculations Result of Researcher)

Table 4. Recap of Organization Contect Clause

Clausul	Descriptions	(%)
4.1	understand the organization and its context	88
4.2	determine the scope of the quality management system	86
4.3	Quality management systems and processes	88
	Mean	**87**

(Source: Data Calculations Result of Researcher)

5.2 Analysis of Leadership Clause

Based on gap analysis for leadership clauses get value 83%. This shows the working procedure has not been fully implemented. Constraints faced is the focus on improving customer satisfaction has not been maintained well in addition to the quality policy is not fully understood until the operator level. Table 5 is the result of gap analysis in the leadership clause.

Table 5. Recap of Leadership Clause

 https://doi.org/10.1051/shsconf/20184901008

Clausul	Descriptions	(%)
5.1	Leadership and commitment	84
5.2	Policy	86
5.3	Organization role, responsibility and authority	79
	Mean	**83**

(Source: Data Calculations Result of Researcher)

5.3 Analysis of Planning Clause

Planning clause get value 84% indicating that the company has not been consistently running the work procedures that have been made. The constraint encountered is that the actual achievement of the implemented quality management system is not in accordance with the desired plan. Table 6 is a recap of percentage value of planning clause.

Table 6. Recap of Planning Clause

Clausul	Descriptions	(%)
6.1	Action to handle risk and opportunities	84
6.2	Quality goals and planning to achieve them	84
6.3	Planning changes	85
	Mean	**84**

(Source: Data Calculations Result of Researcher)

5.4 Analysis of Support Clause

The support clause get value 82%. This value is the same as with previous clause result that the co mpany has not consistently run the working procedure. The constrains faced in clause 7.2 is the worker's competence to job specialization has not been well defined by the company. Besides that quality policy, quality objectives are not well understood by the workers. Table 7 is a recap of percentage value of support clause.

Table 7. Recap of Support Clause

Clausul	Descriptions	(%)
7.1	resource	**82**
7.2	competency	**83**
7.3	awareness	**83**
7.4	communication	**82**
7.5	documented information	**82**
	Mean	**82**

(Source: Data Calculations Result of Researcher)

5.5 Analysis of Operational Clause

The operational clause obtain a gap 94%. This means that many work processes and requirements are run but not consistent. Table 8 is the result of gap analysis performed at the operational clause.

Table 8. Recap of Operational Clause

Clausul	Descriptions	(%)
8.1	Planning and operational control	94
8.2	requirements products and services	94
8.3	external products control and services provided	94
8.4	production and service provision	95
8.5	release of products and services	94
8.6	constraints on unsuitable output	95
	Mean	94

(Source: Data Calculations Result of Researcher)

5.6 Analysis of Job Evaluation Clause

The evaluation clause get value 93%. Indicating that work procedures and requirements are executed but not yet consistent. Table 9 is the result of work evaluation clause.

Table 9. Recap of Work Evaluation Clause

Clausul	Descriptions	(%)
9.1	pemantuan, pengukuran, analisis dan evaluasi	92
9.2	audit internal	93
9.3	Tinjauan Manajemen	94
	Mean	93

(Source: Data Calculations Result of Researcher)

5.7 Analysis of Improvement Clause

The improvement clause get value 88%. This means companies still ignore some working procedures that have been made so that still need to be repaired. The constrains faced in clause 10.1 where in case of incompatibilty, the company's follow up to take action is still not maximal so often found complaints from customers in the same thing. Below table is the result of improvement clause percentage.

Table 10. Recap of Improvement Clause

Clausul	Descriptions	(%)
10.1	general	88
10.2	nonconformities and corrective actions	88
	Mean	88

(Source: Data Calculations Result of Researcher)

5.8 Analysis of Risk Management

Based on interviews with operational managers and some managers, some risks may be found in the company. The risk are mixing of knitting material on tiplash process. Chemical radiation on dye process, process error on packaging departement. The company has actaully set rules to always use personal protective equipment for dyeing process., but the awareness of operators to use self-protective equipment that has been provides is still low. Working prosedures have also been attached to each part of the work, but the accuracy of worker who are still lacking.

6. Conclusion

The results of overall gap analysis can be concluded that the effectiveness of ISO 9001: 2015 implementation is in the range 82% - 94%. This means that there are some work procedures not yet executed but other requirements run well. Constraints facing the company in running the quality management system ISO 9001: 2015 are less control from management representative regarding quality management system implementastion (Clause 4), focus on improving customer satisfaction has not been maintained well besides quality policy not fully understood until operator level (clause 5), implements the quality management system, the desired outcomes are not in accordance with the target (clause 6), worker's competence to job specialization has not been well defined by the company. Besides the quality policy, quality objectives are not well understood by workers (clause 7), if there is a discrepancy, follow-up company to take action is still not maximal so often found complaints from customers in the same (clause 10).

References

1. Prakasa (2016). *Analysis of the implementation of ISO 9001: 2008 quality management system using Gap Analysis*.Universitas Brawijaya.Vol. 3 No. 1
2. Feigenbaum, A. V.(1983).*Total Quality Control*, Third Edition. Terjemahan Ir. Hudaya Kandahjaya, M.S. Jakarta: Erlangga
3. Psosmas, E. (2014). Performance measures of ISO 9001 certified and noncertified manufacturing companies. *Benchmarking: An International Journal, 21*(2014), 756-774
4. Purwanggono, Bambang & Adi Purwadi.(2017).*Analysis of ISO 9001: 2015 certification readiness PT Wijaya Naga Tsupazki by Using Gap Analysis Method*. J@TI. Universitas Diponegoro Vol. IV. No. 3
5. Tukiran, M. (2016).*Building Quality Management System Based on ISO 9001: 2015*. Yogyakarta: Leutikaprio Nauvaliter.
6. Aksorn, Thanet & Hadikusumo, B. H. W. (2007). Gap Analysis Approach for Construction Safety Program Improvement.*Journal of Construction in Developing Countries*.Vol.12,No. 1.
7. Admaja, Awangga Febian Surya. (2013). Preparedness Study of Directorate of Standardization in Implementing SNI ISO / IEC 17065. *Buletin Pos dan Telekomunikasi*. Vol. 11, No.3.

Analysis of OHSAS 18001: 2007 standard renewal towards ISO 45001: 2018 at PT. Power Plant Indonesia by using gap analysis method

Ramy Yahya[1,*], *Naniek* Utami Handayani[1] and *Bambang* Purwanggono[1]
[1]Industrial Engineering Department, Faculty of Engineering, Diponegoro University, Jl. Prof. Soedarto, Tembalang, Semarang, Indonesia 50275

Abstract - The research aimed to understand the current condition of PT. Power Plant Indonesia by evaluating its OHSAS 18001:2007 certification and its preparedness for ISO 45001:2018 certification. Method used for data analysis was gap analysis. Calculation in gap analysis used a checklist compiled according to the requirements listed in OHSAS 18001:2007. Results of the checklist showed that the general requirements clause averaged at 93.88%, health & safety policy clause at 93.97%, planning clause at 92.91%, implementation & operation clause at 91.85%, checking clause at 92.32%, management review clause at 91.85%, and overall score measured against OHSAS 18001:2007 at 92.85%. With the similarities between the points of clauses in OHSAS 18001:2007 and in ISO 45001, the large overall score when measured against OHSAS 18001:2007 signified that the company was prepared for ISO 45001:2018 certification.

Keywords: OHSAS 18001:2007, ISO 45001:2018, gap analysis

1. Introduction

Free competition in industry due to globalization reminds industries both in manufacturing and in service to understand the importance of quality in products as well as quality system in their manufacture. Other than focusing on quality, recently lots of company tries to increase efficiency to face current global crisis. Occupational Health and Safety (OH&S) is one of the factors of consideration. Good practices in occupational health and safety will increase company productivity. To assure that OH&S practices are enacted, companies can administer occupational health and safety management system (OH&SMS) according to Governmental Regulation Number 50 Year 2012 or to the international standard of Occupational Health and Safety Assessment Series (OHSAS) 18001 [1].

International Labor Organization (ILO) (2013) stated that 2.34 million workers died due to their occupation. 2 million of those deaths were related to health and the rest of them were related to work accident. According to Institute of Occupational Safety and Health (IOSH) (2013), an estimated number of 660,000 workers died yearly from cancer arising from work [2].

* Corresponding author : ramyyahya007@gmail.com

Faced with such a condition, ISO 45001 was developed so that organizations or industries could manage OH&S risk and improve OH&S performance. ISO 45001 is the new international standard for occupational health management, replacing OHSAS 18001 [3].

Organizations that observe OHSAS 18001 delegates the responsibility for occupational health and safety to OH&S manager instead of integrating OH&S to the organization operation system. On the other hand, ISO 45001 demands the integration of occupational health and safety into the overall organization management system, hence encouraging top management to play a strong role in the leadership of OH&S management system.

PT. Power Plant Indonesia is one of the companies that has complied with OHSAS 18001. But, according to The Auditor (2016), International Standardization Organization (ISO) will publish in March 2018 ISO 45001 about Occupational Health and Safety Management System (OH&SMS) to replace OHSAS 18001. With the upcoming publication in mind, PT. Indonesia Power, which has complied with OHSAS 18001 with a still valid certification, had a three-year period up to 2021 to complete its migration to ISO 45001.

Therefore, the research aimed to analyze the current condition of the company and compare it to the necessary condition according to the standard of ISO 45001 using gap analysis method. The research wished to obtain result in the form of comparison between the existing system and the necessary system so disparities between the two could be identified and improvements could be made to assist the company in the change towards the new standard of ISO 45001.

2. Literatur Review

2.1. OHSAS 18001 and ISO 45001

OHSAS 18001 standard has been developed since 1999. The framework was developed by a number of standardization and certification bodies as a response to the pressing need of industries for a standard in occupational health and safety management that is measurable and verifiable. The basic aim of OHSAS 18001 was to support and promote good practices in occupational health and safety through systematical and structured management system. Thus, certified organizations could stay in a secure position due to minimization of work accident risks in employees and other affected parties. OHSAS stated requirements for OH&S management system, so the organization could manage risks in OH&S and improve performance. OHSAS standard focused on managing occupational health and safety aspect, not on regulating other areas in health and safety such as employee health/welfare program, product safety, property damage, or environment impact.

OHSAS 18001 (2007) has the following elements, such as: (a) related to the general requirements, organization/industry should make, document, preserve, and sustainably improve OHSAS 18001; (b) devise OH&S policy; (c) devise plan, which consists of: (1) conducting hazard identification, risk assessment, and control determination, (2) reviewing legal and other requirements relevant to OHSAS 18001, and (3) determination of objectives and program of OHSAS 18001; (d) devise application and operation process, which consists of: (1) resources, roles, responsibility, accountability, and authority, (2) competence, training, and awareness, (3) communication, participation, and consultation, (4) documentation, (5) control of documents, (6) operational control, and (7) emergency preparedness; (e) examinations, such as: (1) monitoring and measurement of OHSAS 18001 performance, (2) evaluation of compliance with OHSAS 18001, (3) investigation of incidents, nonconformity, corrective actions, and preventive actions, (4) control of records, (5) internal audit; and (f) management review.

There are a few fundamental differences between ISO 45001 and OHSAS 18001, such as: (a) organizations/industries need to state internal and external contexts which influence organizations/industries in OH&SMS, (b) ISO 45001 doesn't require companies to appoint a management representative, (c) when planning activities, ISO 45001 requires organizations/industries to determine OH&S risk and opportunity that influence said organizations/industries, (d) OH&S related risks are defined as risks related to hazards and to legal and other requirements as well as overall organizations/industries overall context and (e) organizations/industries shall plan actions to handle risks in terms of OH&S threats and opportunities and evaluate their effectiveness.

According to Jones (n.d.), the difference between OHSAS 18001 and ISO 45001 has implications for organizations/industries looking for certification and consultation as well as auditors auditing the new standard. This implication includes the requirement to show adequate leadership and deliberation in the context of organizations/industries. For example, auditor might need to gain new skills, be they interpersonal, in dealing with top management, and technical, in dealing with the proliferation of many information sources and in analyzing the context of the organization/industry. OH&S professionals might need to develop or improve their skills in helping managers to evaluate internal and external factors that influence context, assure an effective involvement alongside with relevant stakeholders and interests, and prepare CEO and other senior managers for audits using this new standard. Introduction of ISO 45001 also allows for a stronger work relation among OH&S professionals and also a more visible and more socially responsible leadership for Board/CEOs [4].

Table 1. Draft of ISO 45001: 2018

ISO 45001:2018			
0.	Introduction	7.	Support
1.	Scope	7.1	Resources
2.	Normative references	7.2	Competence
3.	Terms and definitions	7.3	Awareness
4.	Context of the organization	7.4	Communication
4.1	Understanding the organization	7.4.1	General
	and its context	7.4.2	Internal communication
4.2	Understanding the needs and	7.4.3	External communication
	expectations of interested parties	7.5	Documented information
4.3	Determining the scope of OH&S	7.5.1	General
	management system	7.5.2	Creating and updating
4.4	OH&S management system	7.5.3	Control of documented information
5.	Leadership and participation	8.	Operation
5.1	Leadership and commitment	8.1	Operational planning and control
5.2	OH&S Policy	8.1.1	General
5.3	Organization roles, responsibilities and authorities	8.1.2	Eliminating hazards and reducing OH&S risks
5.4	Consultation and participation of works	8.1.3	Management of change
6.	Planning	8.1.4	Procurement
6.1	Actions to address risks and	8.1.4.1	General
	Opportunities	8.1.4.2	Contractors
6.1.1	General	8.1.4.3	Outsourcing

6.1.2	Hazard identification and assessment of OH&S risks	8.2	Emergency preparedness and Response
6.1.2.1	Hazard identification	9.	Performance evaluation
6.1.2.2	Assessment of OH&S risks and other risks to the OH&S management system	9.1	Monitoring, measuring, analysis and Evaluation
6.1.2.3	Assessment of OH&S opportunities and other opportunities to the OH&S management system	9.1.1	General
		9.1.2	Evaluation of compliance
		9.2	Internal audit
6.1.3	Determination of legal requirements and other requirements	9.2.1	General
		9.2.2	Internal audit program
6.1.4	Planning action	9.3	Management review
6.2	OH&S objectives and plan to achieve them	10.	Improvement
		10.1	General
6.2.1	OH&S objectives	10.2	Nonconformity and corrective action
6.2.2	Plan to achieve OH&S objectives	10.3	Continual improvement

2.2. Gap analysis

Gap analysis as defined by IT Infrastructure Library (ITIL) is an activity that compares two types of data and identifies their difference. Usually, gap analysis is used to compare two sets of requirements. Gap analysis is often structured around one set of area, topic, or category, making gap analysis efficient in identifying which sector or aspect needs repairing. Gap analysis becomes effective because the checklist compiled is structured and matches the topic. Checklist covers all requirements and is made hierarchically in its review. The checklist will cover general questions and give portrayal of the evaluated topic or category. Questions in checklist are compiled comprehensively and in detail, evaluating the individual questions if necessary. Every question is related to another question to ensure traceability. Here are the steps for gap analysis [5]:

- Score Determination

Scoring system used throughout gap analysis is shown in the Table 2 below:

Table 2 Gap Analysis Score

Score	**Definition**
1	If the organization or company does not understand what is required and does not do so.
2	If the organization or company understands the importance of the activity but does not do so.
3	If the organization or company has documents but has not applied them or if the organization or company has applied them but did not recorded them.

4	If the organization or company is engaged in an activity but inconsistently.
5	If the organization or company is doing well (activity is done consistently).

- Checklist Assessment

Checklist assessment is conducted by respondents according to the current condition of the organization. Respondents selected had the adequate competence. The assessment was done with scoring system according to Table 2.

- Gap Assessment

Gap assessment aimed to recognize how large a gap there was in the company. Percentage value was obtained by summing the scores of each variable and dividing it by the maximum value in the variable. The smaller the gap, the better. To measure preparedness, the obtained value of percentage corresponds to the preparedness of company in implementation as shown in the Table 3 below:

Table 3. Gap Analysis Range

Percentage	Description
75% - 100%	The organization is ready to complete QMS ISO 45001: 2016 and certify.
50% - 74%	The organization still has to improve QMS for the preparation of ISO 45001: 2016
1% - 49%	The organization's QMS is in urgent need of improvement as it differs far from the prerequisites for ISO 45001: 2016.

3. Research Methodology

3.1. Research Methodology

Method used in the research was gap analysis method. Gap analysis was done by performing a comparison between the current condition and the necessary condition to obtain ISO 45001:2016. Used in the comparison were the documents of OS&H management system made by the company. Documents was obtained via interview with a number of managers using OHSAS 18001:2007 internal audit checklist, ISO 45001:2016 internal audit checklist preview, and direct observation at PT. Power Plant Indonesia. Results obtained were assigned scores in the range of 1—5 according to their respective percentage values and the range for of prepared the company was in implementation was made. The range was obtained through discussions with experts. Obtained data would be used to evaluate what gaps there were in the implementation of OHSAS 18001:2007 and ISO 45001:2016. The gaps would be sorted by their priorities for improvement, exhibited by whichever gap had the most percentage value [6].

3.2. Research Variables Identification

Research variables were derived from the clauses of OHSAS 18001:2007, as shown as in Table 4 below:

Table 4. Research Variables

No	Clause
1	General Requirements
2	Health & Safety Policy
3	Planning
4	Implementation & Operation
5	Checking
6	Management Review

4. Data Processing

4.1. Data Processing

Data processing in this step was to evaluate the score obtained in Table 4.1. The result of the evaluation was used to measure the preparedness of PT. Power Plant Indonesia in terms of each clauses in OHSAS 18001:2007. Total scores for each variable from the checklist result is shown in the Table 5 below:

Table 5. Score Results from Checklist Data Hasil Skor Data Checklist

Clause	(%)
General Requirements	93.88 %
Health & Safety Policy	93.97 %
Planning	92.91 %
Implementation & Operation	91.85 %
Checking	92.32 %
Management Review	92.16 %
Total Mean	92.85 %

4.2. Analysis of gap analysis results in OHSAS 18001:2007 and comparison with ISO 45001:2018

a. Analysis of Result from General Requirements Clause Checklist

After gap analysis was conducted for General Requirements clause, the value obtained was 93.88%. Table 6 below shows the result of gap analysis performed on the organization context clause.

Table 6. General Requirements Clause Recapitulations

No	ISO 45001	OHSAS 18001	(%)
1	4.4 OH&S management system	4.1 General requirements	93.88%
		Preparedness level	93.88%

b. Analysis of Result from Health & Safety Policy Clause Checklist

For analysis of Health & Safety Policy clause, the value obtained was 93.97%. This value was the gap in the company for the Health & Safety Policy clause in OHSAS 18001, which also appeared in ISO 45001. The table below shows the result of gap analysis performed on Health & Safety Policy clause.

Table 7. Health & Safety Policy Clause Recapitulations

No	ISO 45001	OHSAS 18001	(%)
1	5.2 OHS Policy	4.2 OH&S Policy	93.97%
		Preparedness level	93.97%

c. Analysis of Results from Planning Clause Checklist

Furthermore, for the planning clause, a number of values were obtained from the three clauses with an overall average preparedness value of 92.91%. The value signified that the company was ready for migration. This was supported by the corresponding clauses in ISO 45001. The table below is the result of the gap analysis performed on the Planning context clause.

Table 8. Planning Clause Recapitulations

No	ISO 45001	OHSAS 18001	(%)
1	6.1.2 Hazard identification and assessment of OH&S risks	4.3.1 Hazard identification, risk assessment and control determination	92.94%
2	6.1.3 Determination of legal requirements and other requirements	4.3.2 Legal & Other requirements	92.35%
3	6.2.1 OH&S objectives	4.3.3 Objectives and Programs	93.44%
		Preparedness level	92.91%

d. Analysis of Results from Implementation & Operation Clause Checklist

In the result of the OSHAS 18001 checklist, the Implementation & Operation clause consisted of 7 clauses for which equivalent clauses existed in ISO 45001. The table below details the scores for each clause in Implementation & Operation clause.

Table 9. Implementation & Operation Clause Recapitulations

No	ISO 45001	OHSAS 18001	(%)
1	7.1 Resources	4.4.1 Resources, roles, responsibility, accountability and authority	93.52 %
2	7.2 Competence	4.4.2 Competence, training and awareness	89.91 %
3	7.3 Awareness	4.4.3 Communication, participation and consultation	92.64 %
4	7.5.1 General	4.4.4 Documentation	90 %
5	7.5.2 Creating and updating	4.4.5 Control of documents	91.17 %
6	8.1 Operational planning and control	4.4.6 Operational control	92.94 %
7	8.6 Emergency preparedness and response	4.4.7 Emergency preparedness and response	92.80 %
		Preparedness level	91.85 %

e. Analysis of Results from Checking Clause Checklist

In the results, for the checking clause of OHSAS 18001, there existed 5 clauses, many of which corresponded to the clauses in ISO 45001. The result is detailed in Table 10 below.

Table 10. Checking Clause Recapitulations

No	ISO 45001	OHSAS 18001	(%)
1	9.1 Monitoring, measurement, analysis and evaluation	4.5.1 Performance measurement and monitoring	92.35 %
2	9.1.2 Evaluation of compliance with legal requirements	4.5.2 Evaluation of compliance	91.17 %
3	10.1 Incident, nonconformity and corrective action	4.5.3 Incident investigation, non-conformity, corrective action and preventive action	94.58 %
4	7.5.3 Control of document information	4.5.4 Control of records	91.37 %
5	9.2 Internal Audit	4.5.5 Internal Audit	92.15 %
		Preparedness level	92.32 %

f. Analysis of Results for Management Review Clause

Lastly, the OHSAS 18001 checklist for management review clause resulted in preparedness value of 92.16%.

Table 11. Management Review Clause Recapitulations

No	OHSAS 18001	(%)
	4.6 Management review	92.16 %
	Preparedness level	92.16 %

4.3. Document Draft on Occupational Health and Safety Management System (OH&SMS) Suggested for Migration to ISO 45001:2018

The document on Occupational Health and Safety Management System which is suggested to the company is one of the mandatory document forms in ISO 45001:2018 and is according to the requirements contained in the ISO 45001:2018 clauses. The document draft is expected to assist PT. Power Plant Indonesian in preparing to migrate from OHSAS 18001:2007 certification to ISO 45001:2018. The most common document details used on the implementation of Occupational Health and Safety Management System (OH&SMS) is provided below in Table 12.

Table 12. Mandatory Documents Required by ISO 45001: 2018

No	Mandatory documents	Clause
1	Scope of the OH&S management system	4.3
2	OH&S Policy	5.2
3	Scope of the OH&S management system	5.3
4	OH&S process for addressing risks and opportunities	6.1.1
5	Processes needed to address risks and	6.1.1

	opportunities	
6	Methodology and criteria for assessment of OH&S risks	6.1.2.2
7	OH&S objectives and plans for achieving them	6.2.2
8	Communication	7.4
9	Operational controls	8.1.1
10	Emergency preparedness and response process	8.2

5. Conclusions

According to the research, conclusion can be drawn from the results of gap analysis, using standard checklist of OHSAS 18001 which contains a few similarities to the newest standard of ISO 45001. Overall, PT. Power Plant Indonesia preparedness for the certification scored at 92.85%. Then, reviewed from the preparedness in the individual clauses, the clauses of general requirements, health & safety policy, planning, implementation & operation, checking, and management review, all had a high average score of preparedness. In each of the points in the standard clauses of OHSAS 18001, there existed numerous similarities to the points in ISO 45001. This signified that PT. Power Plant Indonesia was prepared to migrate to the certification according to ISO 45001 standard.

References

1. BSI OHSAS 18001: 2007
2. DIS. ISO 45001, health and safety management systems, 2017
3. ILO. The prevention of occupational diseases. Switzerland: ILO, 2013.
4. Jones, R. (n.d.), ISO 45001 and the evolution of occupational health and safety management systems
5. M. Picard., A. Renault, B. Barafort, and S. "Cortina Measuring readiness for compliance: A gap analysis tool to complete the TIPA process assessment framework: Springer international publishing Switzerland, 633(2016), 106 – 116, (2016).
6. R. Kleinovà., P. Szaryszovà, The new health and safety standard ISO 45001:2016 and its planned changes, (2014).

Analysis of ISO 14001 implementation and program performance assessment ratings company (PROPER) environmental management in. Power Plants with gap analysis tools

Esaul Wairon [1,*], *Bambang* Purwanggono[1], *Naniek Utami* Handayani[1]

[1]Industrial Engineering Department, Faculty of Engineering, Diponegoro University, Jl. Prof. Soedarto, Tembalang, Semarang, Indonesia

Abstract. In various studies revealed that the ISO 14001 Environmental Management System into an integrated environmental management system that is considered capable of improving environmental performance and financial performance. This study aims to analyze the application of ISO 14001 Environmental Management System certification on environmental performance and company performance. The sample in this study was one of the power plants in Indonesia that was ISO 14001 certified in 2015. This study used annual data from the period 2013 to 2014. The main method used was the Gap Analysis Tool. , The results using a checklist show that, in general, the clause requirements have an average value of 50% - 74% the organization still has to fix the SML for preparation of ISO 14001-2015 the inspection process and the clause have an average value, 32% clause, the planning clause has an average value of 77%, the management review has an average value 57% and the overall average value in accordance with ISO 14001-2015 requirements is 55%. based on the characteristics of the sample company, the company needs to improve its inspection and management review so that it can improve the operational management of the environment better.
Keywords : Implementation of ISO 14001, GAP Analysis, Data Proper.

1. Introduction

Indonesia is a developing country with a high industrial activity sepertti Jakarta, Tangerang, Surabaya, Pekanbaru and Bandung have a concentration of pollutants SO2, PM10, PM2,5, which is very high compared to other cities Exposure to pollution can cause harm in terms of both comfort and economy due to disruption of public health, even to result in death (Z. Aprilasani, C. A. Said, T. EB et al ,. Nakazawa et al) [1]." There needs to be awareness of the businesses on the importance of the interaction between the economy and the

* Corresponding author: essaulwairon123456@gmail.com

environment, as well as the transition from the familiar neoclassical economics have come to understand ecological economy that takes into account the environment and ecosystem services, so that sustainable development can be achieved (Perrings, 1987; Common and Stagl, 2005; Miller and Spoolman, 2016)[2].", ISO 14001 Environmental Management System (EMS ISO 14001) issued by the International Organization for Standardization (ISO) in 1996 in Geneva, Switzerland. This system is believed to help create an integrated mechanism in improving environmental performance on an ongoing basis is applied to the activities of daily production (Kitazawa and Sarkis, 2000)[3]." The problem that exists is that not all companies are willing and able to implement the EMS ISO 14001. In addition to the voluntary, some studies reveal that the ISO 14001 EMS certification entails substantial costs depend on the characteristics and facilities of each company which includes investment costs and the cost of regular audits (Bansal and Bogner, 2002)[4]." In 2015, more than 300,000 companies in the world certified ISO 14001 EMS. The high participation in the implementation of EMS ISO 14001 is believed to be due to SML ISO 14001 provides benefits such as: the cost efficiency and resources, expand market opportunities, improve their reputation and profitability, reduce coercive government, avoid conflict and improve stakeholder satisfaction Research has been done about the effect of EMS ISO 14001 on environmental performance improvements have been carried out such as: reduction of emissions and waste, 3R, konvservasi water and energy, environmental risk reduction in the energy and gas companies surveyed by (Morrow and Rondinelli 2012) [5]."; reduction of emissions in the coatings industry in China were invented by (Zeng et al, J. Fresner 1988)[6]." reduction in SO2, and NO2 and dust in the cement industry in Vietnam expressed by (Q. Anh Nguyen, L. Hens 2015)[7]." based on the performance of shares in companies in the United States. Research related to the effect of EMS ISO 14001 and the company's performance in Indonesia, among others carried out by (S. Handayani.,2012)[8]." found that the SML ISO 14001 helps companies achieve better PROPER; (M.Sueb, M. Nety and I. Keraf.,2015)[9]." observed that the application of ISO 14001 EMS improve the performance of wastewater treatment facilities in Semarang; (C. G Old, Banjarnahor 2014)[10]." found an increase PROPER as well, The target pressing business that pollute the environment. In the development of command and control approach that is characterized by supervision to ensure compliance with the rules has been disputed effectiveness. There has been a change in the understanding of the characteristics of the environmental challenges facing the world today. The problem of waste disposal and emissions on a large scale have successfully controlled or regulations for the control was complete. The focus of the problem has moved from the issue of obedience to the issue of continuous improvement (continuous improvement), resource use, energy efficiency, and product quality, Legitimacy theory reveals that the company has a contract with the public to conduct business activities based on the values of justice, and how companies respond to various interest groups to legitimize the actions of the company (Sayekti, Wondabio,.A. Prastiwi,2009)[11].".

2. Literature

2.1 Elements of PROPER

Measurement of environmental performance by PROPER. Selection for PROPER PROPER can measure the extent to which companies improve the quality of the environment through

 https://doi.org/10.1051/shsconf/20184901010

the implementation of an environmental management system Evaluation PROPER announced by the Ministry of Environment (MoE) in report form PROPER with symbols of color, namely (a) gold as the best ratings for the company dutifully consistent run production processes are environmentally friendly and ethical and accountable to the public, (b) green for companies with better environmental management than required, (c) the blue to companies that have made efforts to manage the environment, (d) the red for the company that is not in accordance with the regulations, and (e) in black for companies that violate the rules and negligence act to the detriment of the environment (PROPER, 2015). PROPER assessment criteria consists of two categories, namely: (1) the assessment criteria of obedience, related to environmental documents and reporting, control of marine pollution as well as air and water, management of hazardous and toxic substances (B3), and the potential for damage to land. (2) more than the required criteria (beyond compliance), which adjusted is more dynamic with technological developments and global environmental issues. Aspects assessed include SML application, energy efficiency measures and reduced emissions, 3R efforts B3 and non B3 waste, water conservation and reduction of pollutant load of wastewater, protection of biodiversity, and community development programs,Living environment. PROPER performance ratings generally divided into 5 colors with the following definitions, as shown in Table 1.

Table 1. Color table Rank

Color Indicators	Color Explanation
GOLD	Has undertaken environmental management more than required and has undertaken 3R (Reduce, Reuse, Recycle) efforts, implemented sustainable environmental management systems, and made useful efforts for the benefit of society in the long term.
GREEN	Has done environmental management more than required, has had environmental management system, has good relationship with society, including make effort of 3R (Reduce, Reuse, Recycle).
BLUE	Has undertaken the required environmental management efforts in accordance with applicable rules or regulations.
RED	Conducting environmental management efforts, but only partially achieve results in accordance with the requirements set forth in legislation.
Black	Have not undertaken environmental management efforts meaning, intentionally not doing the environmental management effort as required, and potentially pollute the environment.

(Source: Proper Documents Publications, 2014)

Assessment of performance compliance for the companies in the PROPER use of color indicators, starting from the gold, as the best ratings, followed by green, blue, red, and for the worst rank is indicated with black. It is intended to facilitate the public to know the existing ratings. Aspects of assessment in PROPER focused on the assessment of companies' compliance in water pollution control, air pollution control, management of hazardous wastes and toxic (B3), other liabilities related to environmental impact assessment (EIA), the establishment of an Environmental Management System (EMS), conservation and resource utilization, as well as the social activities of the company. (Rakhiemah & Agustia, 2009)[12]."

2.2 Elements of ISO 14001-2015

ISO 14001 was developed from the concept of Total Quality Management (TQM) principled on the activity of the PDCA (Plan - Do - Check - Action), periodically Assessing the overall environmental management system to ensure its continuing suitability, adequacy, effectiveness of environmental management systems to the changes. In principle, the six principles of ISO 14001 - Environmental Management System above can be divided into 17 elements.

Table 2. Principles and Elements of ISO 14001-2015

Model stages ISO 14001	Elements of ISO 14001
4.2 Environmental policy	4.2.1 Environmental policy
4.3 Plenning	4.3.1 Environmental aspects 4.3.2 Requirements and other legislation 4.3.3 Goals and programs
4.4 Implementation and Operation	4.4.1 The role of responsibility resources, customer 4.4.2 Competence of training and awareness
	4.4.3 Communication
	4.4.4 Documentation
	4.4.5 Document control
	4.4.6 Operational control
	4.4.7 Alertness and emergency response
4.5 Examination	4.5.1 Monitoring and measurement 4.5.2 Evaluation of structuring
	4.5.3 Nonconformity of corrective and preventive measures
	4.5.4 Recording control
	4.5.5 Internal audit
4.6 Management review	4.6.1 Management review

(Source: Documents clause-iso-14001-2015)

2.3 Gap Analysis

Gap Analysis defined by the IT infrastruture Library (ITIL) is an activity that compares two kinds of data that identifies the differences Gap Analysis is generally in a set area, topic or category, thus making efficient Gap Analysis to determine which sectors or fields that need to be made anew. Gap Analysis to be effective as a checklist that created tersruktur and in accordance with the topic. The checklist will cover all terms and conditions be created in the hierarchy in pengkajianya. This will involve general question and provide an overview of the topics or categories that will be judged, questions checklist created a complete, detailed and makes an assessment of each individual requirement another for ensure its flexibility (picard, et al, 2016). Here are the steps to perform Gap Analysis:

2.3.1. Determination Score

Table 3. Analysis score gap

scores	Understanding
1	If the organization or company does not understand what is required and does not do so.
2	If the organization or company does not understand the importance of the activity, then it does not.
3	If the organization or company does not understand the importance of the activity, then it does not.
4	If the organization or company is engaged in an activity but not consistent.
5	If the organization or company is doing well (consistently done).

(Source: Initiatives et al, 2015)

2.3.2 Rate Gap

Gap assessment aims to see how big the gap that exist in the company. Percentage value is obtained by summing the scores per variable. The smaller the gap that exist, the better. To measure the readiness of the percentage value that produced showed the readiness of companies to implement ISO 14001-2015.

Table 4. Range GAP

Score	Understanding
75% - 100%	organisasi are ready to complete SML ISO 14001-2015 and certify
50% - 74%	the organization still has to fix the SML for preparation of ISO 14001-2015
1% - 49%	organization or company is in need of repair far from SML ISO 14002-2015

(Source: purwanggono, 2000)

3. Research Methodology

This study was conducted in March and April 2018 at PT. Generator which is located in the city of Semarang. This study was conducted using a case study on the implementation of an Environmental Management System ISO 14001: 2015 in PT. Generator. the method used is primary data retrieval is done to gather information necessary for consideration in the gap analysis. The primary data is done with a questionnaire as a tool for data collection. The questionnaire posed in this study is a questionnaire taken from the clauses SML 14001; 2015 parameters observed in the Environmental Management System (EMS) 14001: 2015 PT. Generator, which 4.3.Plenning clause 4.5 Examination, 4.6 Management review. Clause includes seven (7) components in the Table 5.

Analysis of the data in this study using Analisys Gap method that aims to determine the extent of the gap / difference between an environmental management system in PT. Generator. this time with the provisions of the environmental management system (EMS) ISO 14001; 2015. Proper measurement of performance data and discussed descriptively by implementing an Environmental Management System based on ISO 14001-2015 EMS criteria. The questionnaire used answer scores range from 0 s / d 2 with the following caption: Score 0 = not correspond to reality has not implemented Score 1 = PT. Generator -

Lyrics not carry out this activity Score = This situation is related to the fact Upon completion of the assessment,
a. Score: '0'. PT. Power does not have EMS implementation, or the implementation of any portion does not have the key elements of SML as the ability to implement all the programs and targets to be met to fulfill the standard requirements SML.
b. Score: '1'.PT. Power has had SML implementation, but not yet fulfilled all the requirements according to standards is below standard ISO 14001; 2015.
c. Score 2 PT. Power has determined and documented most of the implementation of the EMS in accordance with ISO 14001; 2015.

Table 5. Research Variables

No.	Clause
4.5 Examination	4.5.1 Monitoring and measurement
	4.5.2 Evaluation of structuring 4.5.3 Nonconformity of corrective and preventive measures 4.5.4 Recording control 4.5.5 Internal audit
4.6 Management review	4.6.1 Management review
4.3 Plenning	4.3.1 Environmental aspects
	4.3.2 Requirements and other legislation 4.3.3 Goals and programs

(Source: Documents-iso-14001-2015 clause)

Table 6. checklist Score Results

No.	Assessment Components	%	POTENCY
0	DRKPL	16%	32%
1	Environmental management system	60%	90%
	Energy Efficiency	41%	64%
	Reduced Emissions	6%	42%
	Water Conservation	12%	42%
	B3-NB3	0%	47%
	TO-HEART	8%	71%
2	ComDev	0,125	63%

(Source: Documents PROPER)

4. Result and Analysis

4.1 Data Processsing

From the resulting data Environmental Management Performance Summary Document (DRKPL) below criteria will be undertaken and have not been applied seen 414% of the total 720.5% and 133% potency which is owned by PT. The power plant to continue improving environmental monitoring Risk Management.

Table 7. Drop Criteria PROPER

Aspects of the assessment	Max value	Value	potential Value	% implementasi
Energy Efficiency	100%	40.5%	100%	2
Decrease in Emissions	100%	40.5%	63.5%	2
3 R Waste B3	49%	0%	34%	1
3R Non B3 Solid Waste	0%	44%	30%	1
Water Efficiency and Load Decrease	100%	42%	12%	1
Water Efficiency	100%	42%	12%	1
Decrease of Water Pollution Load	99%	42%	5.5%	2
Protection of Biodiversity	12.5%	0%	62.5%	1
Community empowerment	100%	8%	69%	2
Environmental Management System (EMS)	60%	0%	100%	1
TOTAL Value DRKPL	**720.5%**	**133%**	**414%**	**14 %**

(Source: the calculation results)

In the clause. 4.6 Management review shows that the percentage of Evaluation of structuring reaches 90% in this case the company is able to meet the proper standards in management review and for clause 4.3. planning reaches 71% and Goals and programs reach 97% so the company in this case has a green rating in implementing its environmental management.

Table 8. Checklist Score Results

Clause	%
4.5 Examination	32%
4.6 Management review	57%
4.3 Plenning	77%
Total Mean	55%

Table 9. Analysis of Implementation of ISO 14001 and Operations Recapitulation Clause

Clause	Point Clause	%
4.5 Examination	4.5.1 Monitoring and measurement	32%
4.6 Management review	4.5.2 Evaluation of structuring	90%
	4.5.3 Nonconformity of corrective and preventive measures	64%
	4.5.4 Recording control	42%
	4.5.5 Internal audit	42%
	4.6.1 Management review	47%

	Readiness level	57%
4.3 Plenning	4.3.1 Environmental aspects	71%
	4.3.2 Requirements and other legislation	63%
	4.3.3 Goals and programs	97%
	Readiness level	77%

(Source: the calculation results)

5. Conclusion

Based on research that has been done, then the conclusion can be drawn from the results of the gap analysis, using ISO 14001 satdard checklist in which there are some similarities with the standard of the whole PT. Power Plant Indonesia has the readiness to certify 90% Evaluation of structuring, Then views of the readiness of each clause, ranging from a general requirement clause, health & safety policy, planning, implementation and operation, checking and management review the average value of high readiness. Where in any points of the clause indicates that the company PT. Power Plant Indonesia ready to Maintain Environmental Management.

Acknowledgement

In this Researcher authors would like to thank Dr. Naniek Utami Handayani, ST., MT and Dr Ir Bambang Purwanggono, M. Eng guide you and give good feedback and can be carried out this research. as well as those who have provided input both from outside and from the inside so that it can settle this study.

References

1. Z. Aprilasani, C. A. Said, T. EB et al ,. Nakazawa et al, "*The Effect Sertifikasih Environmental Management System ISO 14001 On performance of the company*" Scientific Journal of Management, Volume VII, (2017).[1]."
2. C. G Old, Banjarnahor, "*Analysis of Application of ISO 14001 Environmental Management System PT.Indonesia Semarang Power*", Thesis, University of Diponegoro in Semarang (2014).
3. Bansal, Bogner. "Deciding on ISO 14001*: Economics, Institutions, and Context*" Long Range Planning, vol 35 (2002).
4. Crocker, Tschirhart, Perrings, Settle et al., Tilman et al., Wätzold et al., "*Economic Valuation of Ecosystems*" (2016).
5. Zeng, J. Fresner "*Cleaner production as a means for effective environmental management*" Journal of Cleaner Production vol 06, page 171-179 (1988).
6. Kitazawa S., and J. Sarkis, "*The relationship between ISO 14001 and continuous source reduction programs*" International Journal of Operations & Production Management, Vol. 20 (2000).
7. C.Comoglio, S. Botta ,. Morrow, Rondinelli "*environmental performances improvement: a survey on ISO 14001 certified companies in the automotive sector*" Journal of Cleaner Production 20 (2012).
8. Q. Anh Nguyen, L. Hens "*Environmental performance of the cement industry in Vietnam: the influence of the ISO 14001 certification*" Journal of Cleaner Production 96 (2015).

9. S. Handayani., *"Management of Industrial Limba" Volume 19 Journal of Management "(*2015).
10. M.Sueb, M. Nety and I. Keraf., *"Relation of ISO 14001 Environmental Management System and Financial Performance"* Vol 3, No 1 (2012).
11. S.wondabio,.A. Prastiwi, *"The Influence of Social Responsibility Corpotare Against the company's financial performance"* Ekonomy journal Diponegoro University (2009).

Analysis of customer satisfaction to quality of service using Importance Performance Analysis (IPA) on PT. Telekomunikasi Indonesia. (Case study Telkom Divre IV Central Java & Yogyakarta)

Arlita Rahma Widyasrini[1,*] and *Nia Budi* Puspitasari[1]
[1]Industrial Engineering Department, Diponegoro University, Semarang, Central Java, Indonesia

Abstract. PT. Telekomunikasi Indonesia, Tbk. (PT.Telkom) always serve to provide the best service to customers. Analysis of company performance based on customer satisfaction with the results of research can be done service improvements that are not in accordance with customer expectations and foster customer loyalty and improve the competitiveness of the company PT. Telkom Divre IV Central Java & Yogyakarta. The number of fast subscribers for the 201X period is 252,946 calls. Total quantity of product in Semarang if compared to quantity, if in the long term can cause customer dissatisfaction which resulted the change of customer to company, hence needed analysis which aim for service quality, service quality service analysis perceived by customer, and analysis the level of customer satisfaction PT. Telkom Divre IV Central Java & Yogyakarta. The method used in measuring consumer satisfaction is the method of Importance Performance Analysis (IPA), comparing the expectation of customers with the performance of the company. The use of IPA methods can identify attributes with poor and existing performance from users, and determine the steps of important attributes that still have poor performance, thus the results of the study provide recommendations to PT. Telkom Divre IV Central Java & Yogyakarta.

1. Introduction

In the global era, manufacturing and service companies must face tough competition with existing companies. The company creates its own advantages to compete with its competitors. The company seeks to improve the services provided to customers.

PT. Telekomunikasi Indonesia, Tbk. (PT.Telkom) is one of the companies engaged in telecommunications services. Competition in the telecommunication service industry is very tight to make PT. Telkom always committed to keep trying to give the best service to the customer. PT. Telkom strives to continuously always improve the quality of its performance in order to exist and excel compared with its competitors. Steps to know the service of PT. Telkom has been in accordance with customer needs that is by looking at the

* Corresponding email: arlitarahma.w@gmail.com

number of complaints. Graphic number of consumer complaints in Semarang about Speedy products that go to PT. Telkom in the period 201X can be seen in Figure 1:

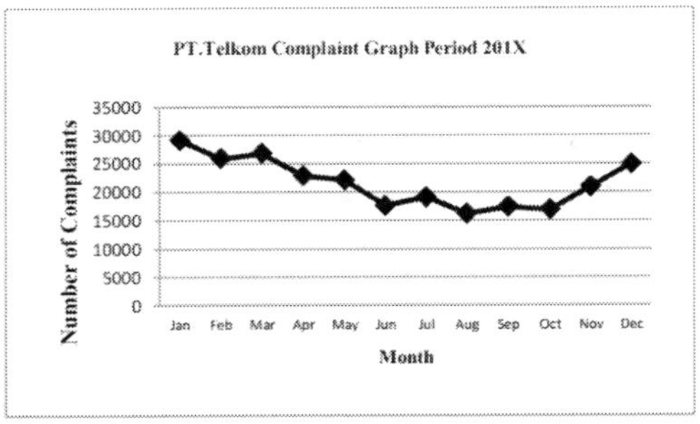

Figure 1. PT.Telkom Complaint Graph Period 201X.

Based on Figure 1 shows that the number of complaints from consumers domiciled Semarang in 201X each month decreased. Overall number of complaints in 201X amounted to 252,946. Consumer complaints data domiciled Semarang about TelkomSpeedy period 201X which has been divided by type of complaint can be seen in Table 1.

Table 1. Formatting sections, subsections and subsubsections.

Complaint Type	**Amou nt**	**Percentag e**
Connections are often disconnected	11.586	4,58%
Slow Speed	14.128	5,59%
Can not browse	22.03	8,71%
Can not connect	185.98 7	73,53%
Modem has not arrived yet	410	0,16%
Etc	18.805	7,43%
Total	252.94 6	100%

Based on table 1. shows one customer can do more than one complaint in one period. Total number of users of TelkomSpeedy products in the city of Semarang, which is 40,596 people. The number of complaints is far more than the number of users of the product when compared in terms of quantity, there are still many consumers who are not satisfied because of the discrepancy of consumer expectations of TelkomSpeedy service. From the number of customer complaints if in the long run occurs continuously may cause customer dissatisfaction resulting in the shift of customers to similar services from other companies.

Improving the quality of service becomes a very important thing for the company to provide value and customer satisfaction. Quality of service has five main dimensions of services that can be used to measure the level of customer satisfaction is tangible, reliability, responsiveness, assurance, empathy [1]. The method used in measuring consumer satisfaction is Importance Performance Analysis (IPA) [2]. The method is used PT. Telkom to compare expectations with what customers receive or company performance. IPA methods can identify attributes with poor performance and complaints from users, and determine the corrective steps of the attributes that are considered important that still have poor performance, so that customers continue to use the services of PT. Telkom Divre IV Central Java & Yogyakarta.

The purpose of this study is to analyze the level of customer expectations of service quality, analyze the performance of service quality perceived by customers, and analyze the level of customer satisfaction PT. Telkom Divre IV Central Java & Yogyakarta.

2. Material and Methods

Research method of customer satisfaction analysis to service quality by using Importance Performance Analysis (IPA) at PT. Telekomunikasi Indonesia, begins with the formulation of problems and research objectives, the research objectives are determined from the formulation of problems that have been identified.

Literature study and preliminary observation. The literature study is conducted to gather as many theories, sciences, data, or information as possible with the analysis of customer satisfaction on service quality using the Importance Performance Analysis (IPA) method. Quality of service centered on efforts to meet the needs and desires of customers and the accuracy of delivery to compensate for customer expectations. There are two main factors that affect the quality of services, the expected service and perceived service. The main dimensions of service quality are arranged in order of their relative importance [3], among others, tangibles, reliability, responsiveness, assurance and empathy.

The Importance Performance Analysis (IPA) method can be started by identifying the initial attributes, identifying the importance level (expectancy) of each attribute and identifying the performance in each attribute. Second, determine the advantages and disadvantages of service with quadrant analysis. Counting the number of incoming questionnaires. Test the reliability and validity of the items with the Microsoft Excel tool. Determining the level of respondent's suitability. Determining the average score of the level of execution / satisfaction and importance level. Determining X is the average of the average score of the level of implementation / satisfaction over all factors or attributes and Y is the average of the importance level score of all factors that affect customer satisfaction. Elaborate the level of these elements into four parts of Cartesian diagram. Cartesian diagram can be seen in Figure 2.

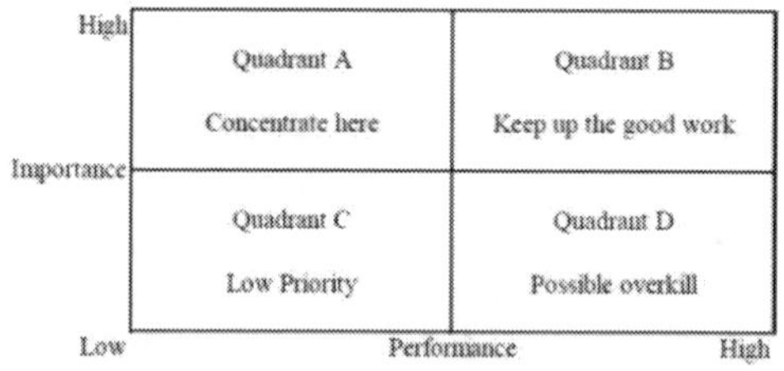

Fig. 2. Cartesian diagram.

Quadrant A shows the service factors or attributes that are considered to affect customer satisfaction including the elements of services that are considered very important, but the management has not implemented in accordance with the wishes of customers to disappointed customers. Quadrant B, indicating the element of service that has been successfully implemented company, for it must be maintained. Considered very important and very satisfying. Quadrant C, shows several factors that are less important to the customer. Implementation by the company mediocre. Considered less important and less satisfactory. Quadrant D, indicating factors that affect customers is less important, but the implementation of the important remains very satisfactory.

Identification of initial research attributes, determination of sample research with Taro Yamane method. One method used to determine the number of samples with known numbers [4].

$$n = \frac{N}{1 + Nd^2} \quad (1)$$

Where n is the number of samples, N is the population number, and d is the margin error.

2.1. Identity Attributes

Identification of attributes is done through literature study by looking for previous studies on the satisfaction of service users of PT. Telkom. From this literature study obtained 24 initial attributes of user satisfaction services PT. Telkom. These initial attributes are used as a basis for the preparation of the questionnaire.

Table 2. Initial Attribute Satisfaction User Service PT. Telkom.

Dimensions	Code	No	Service Attribute	Source
Tangible	T1	1	PT. Telkom uses up-to-date equipment in every work	[5]
	T2	2	The quality of the project result is satisfactory (eg. neat)	[6]
	T3	3	Employees of PT. Telkom wears polite and neat clothes when we work	[5]
	T4	4	Results of work in accordance with the initial agreement	[7]
	T5	5	The office physical facilities are attractive	[7]
	T6	6	The lounge area is comfortable	[7]
Reliability	RE1	7	Employees of PT. Telkom is friendly in providing services	[8]
	RE2	8	Employees of PT. Telkom is friendly in responding to customer complaints	[8]
	RE3	9	Employees of PT. Telkom serves quickly and thoroughly	[8]
	RE4	10	PT.Telkom employees are professional towards the wants and demands of consumers	[8]
	RE5	11	PT. Telkom provides a variety of products that can be made a choice of products for customers	[9]
	RE6	12	PT. Telkom is reliable compared to similar competitors	[5]
Responsive ness	R1	13	Ease in obtaining information about products and services provided	[6]
	R2	14	Employees of PT. Telkom is always willing to help consumers	[5]
	R3	15	PT. Telkom responds to requests and complaints quickly	[5]
Assurance	A1	16	Employees of PT. Telkom can maintain the trust given by consumers	[5]
	A2	17	Customers feel safe to use PT.Telkom products and services	[5]
	A3	18	Employees of PT. Telkom has sufficient knowledge and experience	[8]
	A4	19	Employees of PT. Telkom maintains politeness and friendliness when communicatingwith consumers	[5]
Empathy	E1	20	Employees of PT. Telkom knows the needs that consumers want	[5]
	E2	21	Employees of PT. Telkom is able to provide advice to consumers to choose the type of service in accordance with the needs	[6]

https://doi.org/10.1051/shsconf/20184901011

Dimensions	Code	No	Service Attribute	Source
	E3	22	Company employees are able to answer consumer questions clearly	[6]
	E4	23	The company is able to provide the best solution to consumer complaints	[6]
	E5	24	Employees of PT. Telkom gives a fair service	[8]

2.2. Distribution of Questionnaires

In this study the population in question is the people who use TelkomSpeedy products in the city of Semarang, which amounted to 40,596 people. In sampling there are random sampling and non-random sampling. The sampling method used is the non-random sampling is purposive sampling, because there is a requirement that the respondent is a user who has become a member.

Sampling method used is Taro Yamane method on equation 1 with a margin error of 10% and the population of 40,596 people. Based on the calculation obtained amount of sample as much as 99,75. To facilitate the research, the number of samples to be used is 100 respondents.

3. Results

Based on Cartesian diagram figure 3 that goes into quadrant A is the main priority where the respondents feel the attribute is very important but feel dissatisfied to PT.Telkom performance. The attributes include employees of PT. Telkom is friendly in providing services, PT. Telkom responds to requests and complaints quickly, and employees of PT. Telkom is able to provide advice to consumers to choose the type of service in accordance with the needs.

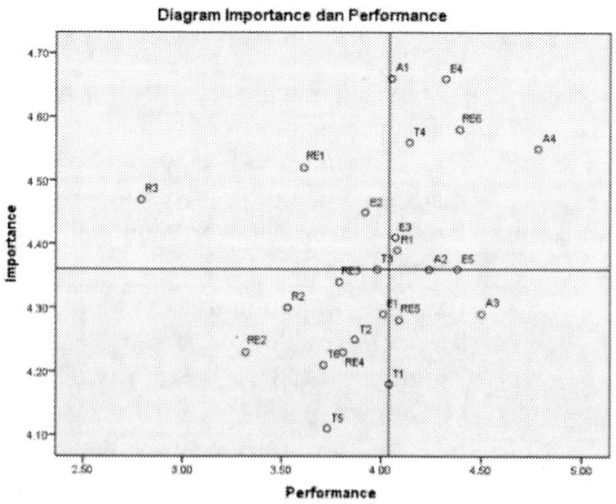

Fig. 3. Diagram of Cartesian Importance Performance.

While in quadrant B users PT.Telkom consider attribute in this quadrant is very important and very satisfactory so that this attribute is obliged to be maintained. these attributes include the results of work in accordance with the initial agreement, PT. Telkom is reliable compared to similar competitors, ease in obtaining information about products and services provided, employees of PT. Telkom can maintain the trust given by consumers, customers feel safe to use products and services PT.Telkom, employees of PT. Telkom maintains civility and friendliness when communicating with consumers, company employees are able to answer consumer questions clearly, and companies are able to provide the best solution to consumer complaints.

In quadrant C shows some less important attributes and less satisfactory, so users behave mediocre, attributes that exist in quadrant C, among others, the quality of the results of the project carried out satisfactory (example: neat), employees of PT. Telkom wearing polite and neat clothes when work, attractive office physical facilities, comfortable waiting room, employees of PT. Telkom is friendly in responding to customer complaints, employees of PT. Telkom serves quickly and thoroughly, PT.Telkom employees are professional towards the wants and demands of consumers, employees of PT. Telkom always willing to help consumers, and employees of PT. Telkom knows the needs that consumers want.

While in the last quadrant is the quadrant D there is no attribute that shows less important factors but the user is very satisfied among others PT. Telkom uses up-to-date equipment in every work, PT. Telkom provides a diverse product that can be a choice of products for customers, employees of PT. Telkom has sufficient knowledge and experience, and employees of PT. Telkom gives a fair service.

Recommendations given to PT. Telkom Divre IV Central Java and Yogyakarta is the first to improve the performance of service quality attributes that are considered important by the user but its performance is still considered low, especially in quadrant A which amounted to 3 attributes (Telkom employees are friendly in providing services, PT Telkom respond to demand and complaints quickly, Telkom employees are able to advise consumers to choose the type of service in accordance with the needs), so that users do not feel disappointed and want to re-use the services of PT. Telkom.

The second recommendation is to maintain and continuously improve the performance of service quality attributes that are considered important and perform well by users, especially in the B quadrant of 8 attributes (Results of work in accordance with the initial agreement, PT Telkom reliable than similar competitors, ease in obtaining information about products and services provided, PT Telkom employees can maintain the trust given by consumers, customers feel safe to use the products and services of PT.Telkom, employees of PT Telkom maintain courtesy and friendliness when communicating with consumers, company employees able to answer consumer questions clearly, the company is able to provide the best solution to consumer complaints), so users feel satisfied and loyal to the services provided.

Acknowledgment

The authors would like to thank Telkom Divre IV Central Java & Yogyakarta and Diponegoro University for support in this research.

References

1. M. A. Fadhillah, *Analisis Kinerja Perusahaan Dilihat Dari Kepuasan Pelanggan Pada Perusahaan Kontraktor Telekomunikasi Dengan Menggunakanpendekatan Metode B2B Servqual Dan Importance Performance Analysis (IPA),* (2012).
2. O. Farreza, *Analisis Kepuasan Konsumen Terhadap Produk Dan Pelayanan Rumah Makan Pecel Lele Lela Bogor,* (2015).
3. J. Goncalves, A. Pinto, M. Batista, A.C. Pereira, G. Ambrosano, *Importance-performance analysis: Revisiting a tool for the evaluation of clinical services*, H, **6**, 285-291. (2014).
4. J. Martilla, J. James, *Importance-Performance Analysis.*, JoM, **41**, 77-79, (1977).

https://doi.org/10.1051/shsconf/20184901011

5. J. Oscar, *Analisis Kepuasan Pelanggan dengan Importance Performance Analysis di SBULaboratory Cibitung PT Sucofindo (PERSERO)*, JTI, 1-10. (2013).
6. A. Parasuraman, V.A. Zeithaml, L.L. Berry, *A Conceptual Model of Service Quality and Its Implications for Future Research,* JoM, **49**, 41-50, (1985).
7. A. S. Singh, M. B. Masuku, *Sampling techniques and determination of sample size inapplied statistics research: an overview,* IJoECM, **11,** 1-22, (2014).
8. M. YuvinaTileng, W. H Utomo, R. Latuperissa *Analysis of Service Quality using Servqual Method and Importance Performance Analysis (IPA) in Population Department, Tomohon City.*, IJoCA, 23-30, (2013).
9. V. A. Zeithaml, , L. L. Berry, and A. Parasuraman, *Communication And Control Process In The Delivery Of Service Quality.*, JoM, **52**, 35-48, (1990).

The urgency of testing standard for elderly's monitoring systems in Indonesia

Sunartoto Gunadi[1], *Zulkalnain* bin Mohd Yussof[2], *Ahmad Yusairi* bin Bani Hashim[3]
[1]Surya University, Physical Energy Engineering,Tangerang15143, Banten, Indonesia
[2]Fakulti Kejuruteraan Elektronik dan Kejuruteraan Komputer, Universiti Teknikal Malaysia, Melaka
[3]Fakulti Kejuruteraan Pembuatan, Universitas Teknikal Malaysia, Melaka

Abstract - Due to several aspects, such as safety, applicability, and reliability of the elderly monitoring systems, it is urgently need a testing standard for this system. Commonly, the user, especially caregivers or nurse, do not know exactly that procedure. In order to educate both, consumer and producer, preventing problems which may arise among them, and guarantee that instrument is operate as well as stated on the instruction manual, which conform to a certain testing standard which introduced by legal institution (SNI or ISO), establishment a testing (physical and/or written) for this device, especially in Indonesia, is urgently needed.
Keywords: testing standard, monitoring systems

1. Introduction

In table 1 belows, USA have elderly people about 15.2% from their population, but for Japan, the number of Japanese people with ages 65 years or older nearly quadrupled in the last forty years, to 33 million in 2014, accounting for 26.3 % of Japan's population.

Table 1. Percentage elderly population in several country [1]

NO	COUNTRY	ELDERLY.S	YEAR	NOTE
1.	JAPAN	33.9 Million	2017	26.3 %
2.	SWEDEN	1.97 Million	2017	19.8 %
3.	UK	11.8 Million	2016	17.8 %
4.	USA	49.2 Million	2016	15.2 %
5.	INDONESIA	21.5 Million	2017	9.43%

Based on statistical data which released by Bureau of Communication and Society Care Ministry of Health, the population of the elderly in Indonesia are more than 9% from total population**[1]**, especially in 3 province, ie Yogyakarta, Middle Java and East Java. It means, according to the definition, Indonesia already belong to ageing country [more than 8 %]. It has to a consequence for the government for introducing more reliable health services.

As stated by Prof. Dr. dr Siti Setiati, SpPD-KGer, Pergemi (Persatuan Gerontologi Medik Indonesia), in paper "Masalah Kesehatan yang Harus diwaspadai oleh Lansia"[2], that there are about 23 % global problem health which belong to Elderly [Lansia] health. It was underlined by Laurie Orlov[3], that for the Elderly, mostly they have at least one of 14 I. In this paper, 14 (fourteen I) are consists of:

1. immobility,
2. instability,
3. incontinence,
4. intellectual impairment,
5. infection ,
6. impairment of vision and hearing, taste, smell, communication, convalescence,
7. integrity,
8. impaction,
9. isolation,
10. inanition,
11. impecunity,
12. iatrogenesis,
13. insomnia,
14. immune deficiency.

Another paper, written by Andrew Carle, executive-in-residence at the Program in Senior Housing Administration at George Mason University, stated that: falls "are also the number one cause of hospitalization due to injury for this age group." That is why, that "Emergency Response Technology especially, elderly monitoring systems, are important, because falls are the number one cause of death due to injury" in people over age 75. In detail, the problem and solutiona is written in paper which topic "Fall Prevention for Older Adults" [4].

2. Recent Condition

Mobility [for people, especially elderly] is represented in the International Classification of Functioning Disability and Health (ICF) by classifier d460 - Moving around in different locations, [see Figure 1], that can be subdivided into:

- d4600 – Moving inside the house;
- d4601 - Move inside buildings than own home;
- d4602 - Move out of your home and other buildings

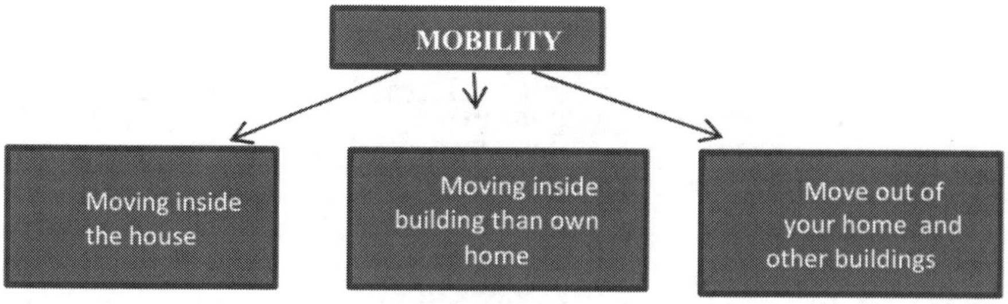

Figure 1. Mobility by ICF

Kirsten K. B. Peetoom and team on their paper *"Literature review on monitoring technologies and their outcomes in independently living elderly people"*[2], stated that there are 5 main groups of monitoring technology, ie:

1. in-home passive infrared motion sensors,
2. body-worn sensors,
3. video monitoring,

4. pressure sensors, and
5. sound recognition.

As every monitoring system have to have an advantages and disadvantages, so using those monitoring systems, it should combine with the type of activity daily living and the mobility of the elderly, which may they choose moving inside the house, or may they move inside buildings than own home or they may move out of your home and other buildings.

3. Problem and discussion

3.1. Problem

In Indonesia, even only 9.43 % the elderly population, but the number mostly 30 million already. In several province, especially in Java, recently, there are a lot of new Private Elderly house, were use some monitoring system. At east Java, a lot of private elderly houses use same monitoring device for their daily activities.

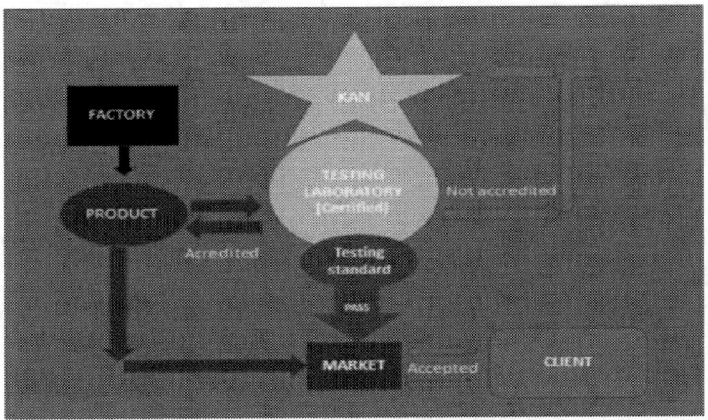

Figure 2. The role of Testing Standard

Due to the problem which may arise, cause by the technical specification of the monitoring system which are not conform to the user applicability or may be the reliability of the system which mismatch with reality, so the problem should be solve by the third party. But, in this case, if the testing standard [SNI or ISO] are available, solution of the problem will be solved easier.

3.2. Discussion

In Indonesia, recently already have about 198 acredited Calibration laboratory and 1043 Test Laboratory [11], unfortunately there is no laboratory which have competence in testing of such kind monitoring system. Relying on transition policy of KAN about ISO/IEC 17025:2017 and considering that Indonesia have to have a testing Laboratory which have competency in such monitoring systems, there are several alternative for overcoming the above problem:

a. The government hand in hand with the private company establish a New Testing Laboratory,
b. To improve established Testing Laboratory, such as LP Electromedic BPFK,
c. The Minister of Research and Technology, appointed Institute of Sciences, in order to developes the competency of both, Research Center for Metrology [P2M]

and Research Center for Quality System and Testing Technology, in overcoming the problem which arise till 2020.

4. Conclusion

From this paper, simply we take a conclusion that:

1. Even recently already start, but anticipating elderly boomer on the next year 2020, it will arise and developed, a new business activity which closed related with elderly people, ie Private Elderly house.
2. As consequences of this Private Elderly house, there are a new demand of equipment which closely related with Elderly daily activity, such as Elderly Monitoring System.
3. To prevent conflict arise about such system which do not perform as the consumer need, earlier Testing Laboratory, which have standard of conformance in testing competency (both, physical and written) is established as earlier as possible.

References

1. World Population Ageing – the United Nations; (2017) *www.un.org/en/development/desa/population/.../pdf/ageing/WPA2015_Report.pdf*
2. S. Setiati, SpPD-KGer, On "Masalah Kesehatan yang Harus diwaspadai oleh Lansia" Pergemi (Persatuan Gerontologi Medik Indonesia), (2016).
3. K K. B. Peetom 1,2, M A. S. L1, M Joore3, C D. Dirksen 3and L. P. De Witte1,21 On Literature review on monitoring technologies and their outcomes in independently living elderly people; Research Centre for Technology in Care, Zuyd University of Applied Sciences, Heerlen, the Netherlands, (2015).
4. "Fall Prevention for Older Adults"; The American Occupational Therapy Association. Inc, (2012).
5. M.R. Marques[1]; T.M. Raymundo[2]; C.S. Santana[3] [1]Master student at Post Graduate Program Interunits Bioengineering EESC/FMRP/IQSC –University of Sao Paulo, Brazil, On Use of sensors systems to monitor the mobility of elderly; Journal of Physics: Conference Series, Volume 477, conference 1 (2013)
6. K. Guan, M. Shao, and S. Wucor; On A Remote Health Monitoring System for the Elderly Based on Smart Home Gateway, (2017).
7. Pedoman Pengembangan Standar Nasional Indonesia; PSN 01, (2007).
8. Population Reference Bureau; Fact Sheet: Aging in the United States, (2016).
9. Office for National Statistics; Overview of the UK population: July 2017, (2017).
10. Aging of Japan; https://en.wikipedia.org/wiki/Aging_of_Japan
11. Kan.or.id – (2018)

Revision of the Japanese Industrial Standardization Act

Ko Ozaki[1]

[1]Standardization Strategy Section, Ministry of Economy, Trade and Industry (METI), Japan

Abstract. METI is, as of the beginning of May 2018, in the process of revising the Japanese Industrial Standardization Act which was enacted in 1949 and has supported Japan's standardization system for almost 70 years. Objectives of the revision are: 1) expansion of the scope into services and other sectors, 2) acceleration of the JIS development process, 3) promotion of international standardization activities, and 4) enhancement of penalties. The new act is expected to come into effect in the latter half of 2019 if it is passed through the Japanese Diet as planned. By making use of the new system, Japan will strengthen both domestic and international standardization activities.

1 Outline of the Japanese Industrial Standardization Act and background of the revision of the act

1.1 Outline of the Japanese Industrial Standardization Act

The Japanese Industrial Standardization Act (JIS Act) [1], which was enacted in 1949 to improve quality and rationalize production of mineral and industrial products, has played a significant role in Japan's standardization. Specifically, it gives legal basis to: the Japanese Industrial Standards Committee (JISC) [2], Japan's national standardization body; the Japanese Industrial Standard (JIS), Japan's national standard; and the JIS mark and Japan National Laboratory Accreditation System (JNLA), Japan's national conformity assessment systems. As of December 2017, there were 10,622 active JIS based on the Act. In addition, as of March 2018, there were 24 JIS-accredited certification bodies, including three outside of Japan, which have issued about 8,700 certifications for JIS Marks [3].

1.2 Background of the revision of the Japanese Industrial Standardization Act

While the importance of the JIS Act has not changed, Japan recognizes that the Act should be revised given the recent trends in standardization activities. First, the scope of the Act, which currently focuses on mineral and industrial products, is too narrow and should be expanded to include services and other sectors in line with international standards. Second, Japan should accelerate the development of JIS given that the development process of international standards, as well as the development of new technologies, has become faster.

Third, while Japan has been active in contributing to and participating in international standardization activities, the current JIS Act primarily focuses on domestic standardization in Japan, and thus increased efforts are needed to put emphasis on international standardization. Finally, in light of several quality management misconducts by Japanese manufacturing companies, Japan is fully aware that it must take measures that can secure credibility for JIS and the Japanese manufacturing industry. To this end, the Japanese government submitted a bill that aims to revise the JIS Act to the Diet on February 27, 2018. [4]

2 Revisions of the Japanese Industrial Standardization Act

2.1 Expansion of the scope into Services and other sectors

The revised act is to newly entitle (digital) data, services, and management systems as targets of standardization under JIS. With regard to JIS for data and management systems, Japan has already undertaken considerable development in this area based on the interpretation of the JIS Act that these JIS are within the scope of "industrial products." However, Japan recognizes that, given that the variety of standards regarding data and management systems have been increasing internationally, it would be difficult to maintain adequate coverage with the current flexible interpretation of the JIS Act. (For e.g., it would be difficult to envision a management system standard specifically for services falling within the scope of standards for industrial products). Thus, Japan plans to clearly define that the JIS Act covers standards for data and management systems. In addition, services standards will be added to the scope of JIS. It will allow Japan to contribute to international discussions of services standards more actively. It should be noted that Japan is also planning to design JIS Mark systems for data and services and stipulate it in the revised act.

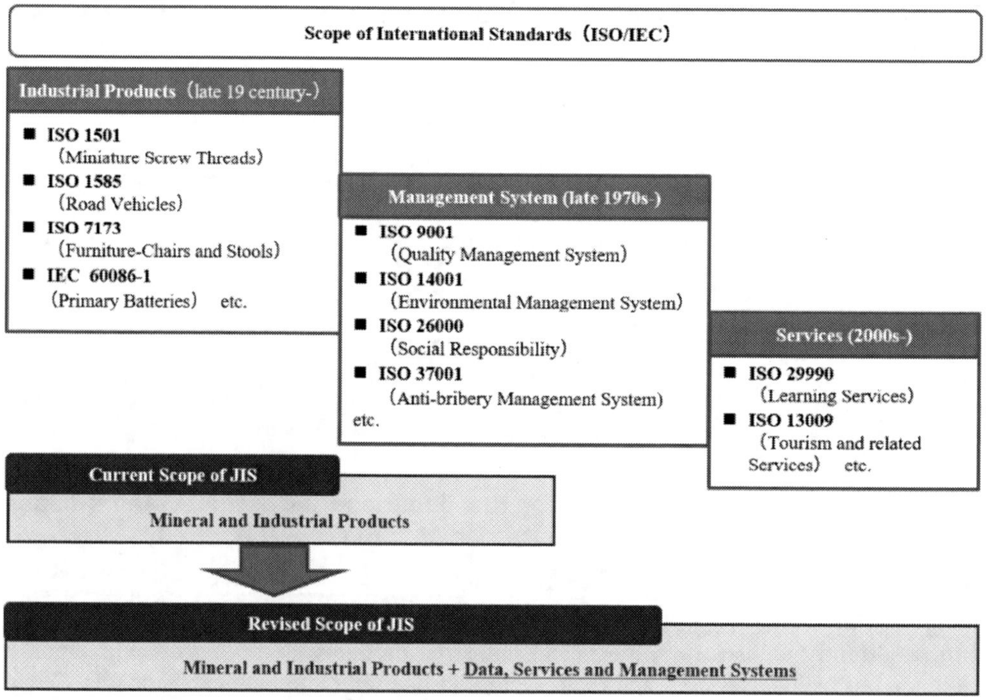

Fig. 1. Scope of International Standards and JIS.

2.2 Acceleration of the JIS development process

The revised act is to allow private associations with expertise to develop JIS without being reviewed by JISC. This new scheme is expected to accelerate the development of JIS. Currently, an individual or organization that aims to develop a JIS needs to form a drafting committee, draft the JIS in the committee, and submit the JIS draft to competent ministers of the Japanese government. Then the competent ministers request JISC to deliberate the draft and, if authorization committees or technical committees under JISC, which comprise of various stakeholders with varied interests in the JIS, consider the draft appropriate, the ministers enact the JIS.

On the other hand, in the new scheme, the deliberation process is entrusted to accredited private entities with expertise, which can be, for instance, industrial associations that have developed a large number of JIS in the past. This means that the accredited private entities are expected to play roles that JISC plays in the current process, such as drafting appropriate JIS while putting diverse opinions from various stakeholders together. Under such condition, JIS drafts written by accredited private entities are not discussed in JISC and competent ministers enact the JIS as soon as the entities submit JIS drafts and the ministers confirm that the drafts are appropriate.

It is expected that, while the period from drafting JIS to enacting it is about two years on average under the current scheme, the period may be shortened under the new scheme. It should be noted, finally, that the current JISC scheme will not be abolished, and stakeholders can select either the current scheme or new scheme based on their preferences.

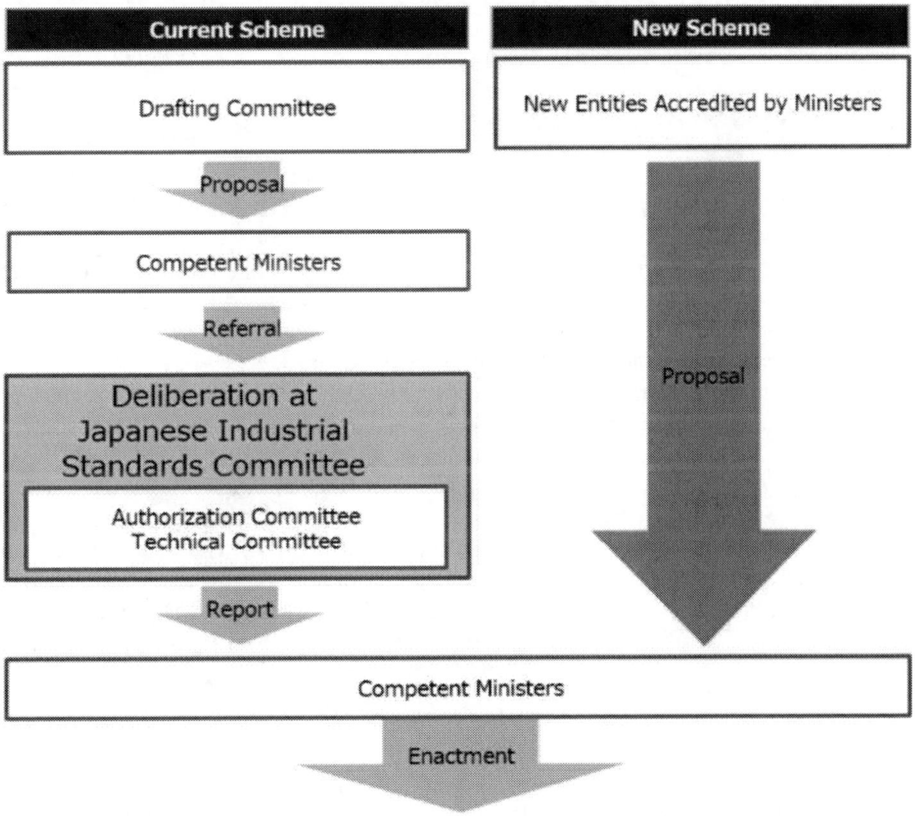

Fig. 2. Acceleration of the development of JIS.

2.3 Promotion of International Standardization Activities

As one of its core purposes, the revised act is to define promotion of international standardization activities. While JISC has actively contributed to international standardization activities in the International Organization for Standardization (ISO) and the International Electrotechnical Commission (IEC), the current JIS Act focuses on national standardization and does not clearly stipulate international standardization. However, given that the significance of international standardization activities for Japan as well as the importance of the harmonization of JIS with international standards have grown significantly, Japan recognizes that it is appropriate to place increased emphasis on the importance of international standardization activities in the JIS Act. Therefore, the revised act is to define promotion of international standardization activities as one of its purposes. Also, it is to stipulate that the government, research institutions, and companies are obligated to make a sincere effort regarding national and international standardization activities.

2.4 Enhancement of Penalties

Along with the three proposals for revision outlined above, the upper amount of the fine for unauthorized labeling of JIS mark is to be increased from the current one million yen to 100 million yen. This is one of the measures that the Japanese government will take in order to respond to recent misconducts, which include breaches of rules for JIS marks, committed by several Japanese manufacturing companies. The Japanese government expects that this measure will mitigate the risk of future misconducts related to JIS and assure credibility of JIS.

3 Making Use of the New System

The new JIS Act is expected to come into effect in the latter half of 2019 if it is passed through the Japanese Diet as planned. While it would take one to two years until the act comes into effect, Japan has already started to work on measures that it can take. For example, JISC proposed international standardization of temperature-controlled refrigerated delivery services to ISO and it has been under discussions in PC 315 [4]. As this example shows, Japan will strengthen its contribution to standardization activities for services as well as international standardization activities in general.

Having said that, Japan has fully recognized that it needs to take measures besides the revision of the JIS Act in order to make the new systems work. For instance, capacity building for experts and organizations that develop service standards could be one of the challenges for Japan. Japan understands that the number of human resources and organizations for the standardization of services is limited partly because such standardization has largely fallen outside the scope of the JIS Act. Thus, Japan aims to take comprehensive measures that include financial support for the development of standards and training programs for experts. Of course, Japan's measures will not be limited to those for standardization of services, and it will take other measures necessary to make the new systems work.

References

1. Ministry of Justice, Industrial Standardization Act http://www.japaneselawtranslation.go.jp/law/detail/?id=20&vm=04&re=02

2. JISC, *Japan and Industrial Standards*, 4-5 (2017)
http://www.jisc.go.jp/eng/policy/Japan_and_Industrial_Standards_2017.pdf

3. METI, Cabinet Decision on the Bill for the Act of Partial Revision of the Unfair Competition Prevention Act, etc. (2018)
http://www.meti.go.jp/english/press/2018/0227_003.html

4. JPO, Promulgation of the Act of Partial Revision of the Unfair Competition Prevention Act, etc. (2018)
https://www.jpo.go.jp/torikumi/kaisei/kaisei2/fuseikyousou_h300530.htm

5. METI, Aratana Kijun Ninsho No Arikata Nitsuite (JAPANESE) (2017)
http://www.meti.go.jp/committee/sankoushin/sangyougijutsu/pdf/006_02_02.pdf

Assesing the Electronic Service Quality using *E-S-Qual* and *Importance Performance Analysis* Combined Method

Naniek Utami Handayani[1]*, *Agus* Tri Wibowo[1], *and Diana* Puspita Sari[1]

[1]Department of Industrial Engineering, Faculty of Engineering, Diponegoro University

Abstract. Internet is an integral part of any service provider activity which serve millions of users with variety of needs throughout the world. The function of internet is widely used and adapted by companies providing services in the form of business transactions, electronically or by means of e - commerce. The quality of service provided is vital to each company's success. This case study was conducted in one of the biggest companies engaged in Indonesia's online shopping business: LAZADA. Assessing the quality of electronic services provided by LAZADA using a combination of E - S - Qual and IPA methods to know the quality of services provided by LAZADA and how customer perceive LAZADA's service. The result of this study procured E - S - Qual value as 2.5 out of a scale of 5, which shows that a lot of things can still be explored by LAZADA in providing services for the customers. Perceptions or expectations of the customers that have not been fulfilled in LAZADA's service delivery are: appropriate duration of delivery, the certainty of the number of goods ready for shipment, and out of stock items.

1. Introduction

In this globalization era, Internet is an integral part of any service provider activity which serve millions of users with variety of needs throughout the world. With more than 3 billion users the Internet became a universal tool, globalizing dimensions worldwide. In addition, the Internet has also become a general and universal tool of information for people, be it at home, school, or work. Within a few years back, the internet become the change to communication and business.

Within the business scope, extremely high market competition conditions as it is now makes the Internet both integral and impactful for companies in various fields of business, particularly in the service providing sector. The consequences or effects of internet use in service-providing sector also impacted on the shift of conventional service providers to internet-based service providers. Those companies grab the opportunity to use electronic transactions and electronic commerce (e-commerce) to interact with consumers/customers.

The main key of success or failure of a company that uses the concept of electronic transactions is not just the web interface offered nor a low price, but rather the electronic service (e - service) quality [1].

The concept of conventional service quality itself has been researched and developed many times. Service Quality itself is one of the critical factors of a successful service provider, because of it is close relation to customer satisfaction. Moreover, the increase of customer satisfaction will cause positive impact on the company's own service providers, such as repurchasing (repeat purchases), customer loyalty and profit [2]. However, e-service quality is different from the conventional-based service quality that includes interaction between the seller and the buyer. The E-S-Qual or Electronic Service Quality is an overall customer assessment of the e-service's delivery in the virtual market [3].

In Indonesia alone, a lot of companies and businesses have utilized Internet to interact with customers. One of them is LAZADA online - shopping. LAZADA itself is one of the largest electronic commerce (e - commerce) in Southeast Asia, reaching 91.4 million dollars in terms of profit during the first 6 months of 2015, equivalent to more than 1 billion IDR (http://www.aitinesia.com),

With a broad marketplace and a huge advantage, theoretically it will have excellent and satisfying service for each customer. High quality service will form customer satisfaction and create customer loyalty [4].

But in reality, the quality of service provided by Lazada is not as expected of one of the largest e - commerce in Indonesia. Based on a review by a company website reviewer, according to comments and reviews as well as ratings given by customers, the quality of service provided by Lazada stands at 1.9 / 5, which categorized as very low (http://trustedcompany.com). One of the reviewer on the website (a customer or prospective buyer of Lazada) wrote Lazada delivery performance as very poor, another potential customer wrote that Lazada disappointingly cancelled order unilaterally.

Customer complaints as written above, if not taken seriously by the company concerned will cause many losses for the company itself. Consumers lack of loyalty, the number of complaints, will lead to lower financial gains.

Therefore, research is needed to assess Lazada's quality of service. Because Lazada itself is an online shopping mall, it can be done using Electronic Service Quality (E-S-Qual) method to assess Lazada online shopping store quality of service.

This study aims to assess the quality of electronic service using a combination method of E-S-Qual and Importance Performance Analysis, it is useful in concluding the expectation of customers of a service contrasted to the reality of service given by Lazada. By using a quadrant graph, the following things will be known: the important but not satisfactory, satisfactory and important, not important but satisfactory, not important and not satisfactory, and unsatisfactory. With the combination of both methods, useful suggestions for improving the quality of service in Lazada will be further provided.

2. Research Methodologies

The E - S - Qual scale or dimension by [5] is used in this study to assess the quality of electronic service, it consists of four dimensions with 22 items statement. The dimension ratings for E - S - Qual is efficiency, system availability, fulfilment, and privacy.

The first dimension of E - S - Qual is efficiency (EFF), defined as the ease and speed of website use or the ease of electronic transaction. In other words, easy to use, neatly arranged, and minimal amount of information needed to be entered by the customer. The second dimension is the System Availability (SYS) which is the technical function rating of the online shopping website. The third is fulfilment as in fulfilment of goods and promises offered to customers. And the fourth is Privacy, namely the security offered by providers of online shopping, for both security and confidentiality of the data entered by customer.

Electronic Service Quality (E - S - Qual) then measured with the following formula 1.

$$E - S - Qual = \sum_{i=1}^{n} WS_{ij} \times PS_{ij} \quad (1)$$

Whereas E - S -Qual equal the value of the item statement, WSij the weight factor of statement J against I, and PSij is the value obtained from i against j statement. The weight factor alone is the result of normalization and justification of importance value which can be formulated at formula 2.

$$WS_{ij} = \frac{IS_{ij} - \min(IS_j)}{max(IS_j) - min(IS_j)} \quad (2)$$

Whereas ISij is the value of the quality of service significance per j statement item against each i.

After that a priority matrix (Importance Performance Analysis) will be depicted to determine the interests and perceptions of each customer to LAZADA, and how the service benefit to these interests.

3. Result and Analysis

The item statement used refer to [5]. Two parts of the questionnaire are used, the first is to assess e-service quality using E - S - Qual method and the second is to assess the importance of each statement itself. In this questionnaire the prerequisite are respondents aged 16 and older that have shopped at online shopping store LAZADA beforehand.

Table 1 Dimensions and Item Statements of E – S – Qual

Dimensions	Item Statements
Efficiency	EFF1: The site makes it easy to find what the customer needs. EFF2: It makes it easy to get anywhere on the site. EFF3: It enables the customer to complete a transaction quickly. EFF4: Information at the site is well organized. EFF5: It loads its pages fast. EFF6: The site is simple to use. EFF7: The site enables the customer to get on to it quickly. EFF8: The site is well organized.
Fulfillment	FUL1: It delivers orders when promised. FUL2: The site makes items available for delivery with-in a suitable period. FUL3: It quickly delivers what the customer order. FUL4: It sends out the items ordered. FUL5: It has in stock the items the company claims to have. FUL6: It is truthful about its offerings. FUL7: It makes accurate promises about delivery of products.
System Availability	SYS1: The site is always available for business. SYS2: The site launches and runs right away. SYS3: The site does not crash. SYS4: Pages at this site do not freeze after the custo-mers enter the order information.
Privacy	PR1: It protect information about the customers' web-shopping behavior. PR2: It does not share the customers' personal informa-tion with other sites. PR3: The site protects information about the customers' credit card.

One hundred and eighty respondents have participated in filling out the questionnaire, consisting employees, students, civil servants, entrepreneurs, and of other positions or occupations. Then a reliability test using Cronbach's Alpha was conducted to check the validity and the correlation value given by respondents for each statement [6]. Cronbach's alpha value of each statement are shown in Table 2. It is known that the value of

Cronbach's alpha for each statement stated as more than 0.7, indicating the questionnaire used as reliable.

Table 2 Cronbach Alpha for Each Dimension of E – S – Qual

Dimension	N of item	Cronbach's Alpha
Efficiency	8	0.923
System Available	4	0.849
Fulfillment	7	0.709
Privacy	3	0.832

The average value of each item then measured with formula (1) and (2). The results of the Electronic Service Quality calculation are shown in the Table 3.

Table 3 E – S – Qual Result

Dimensi	I	W	P	ESQ
EFF1	4.02186	0.75546	4.04372	3.0548837
EFF2	4.13661	0.78415	4.13661	3.2437367
EFF3	4.06557	0.76639	4.06011	3.1116411
EFF4	4.07104	0.76776	4.06557	3.1213831
EFF5	3.97814	0.74454	3.98361	2.9659366
EFF6	4.08197	0.77049	4.08197	3.1451223
EFF7	4.06557	0.76639	4.05464	3.1074532
EFF8	4.03279	0.75820	4.03279	3.0576458
SYS1	4.14208	0.78552	4.14208	3.2536803
SYS2	3.96721	0.74180	3.96721	2.9428917
SYS3	3.66120	0.66530	3.66120	2.4357998
SYS4	3.90164	0.72541	3.90164	2.8302876
FUL1	4.03825	0.75956	1.89071	1.4361134
FUL2	4.14208	0.78552	1.89071	1.4851892
FUL3	3.84153	0.71038	2.11475	1.5022843
FUL4	4.26230	0.81557	1.61202	1.3147227
FUL5	4.23497	0.80874	1.69399	1.3700021
FUL6	4.07650	0.76913	1.57923	1.2146302
FUL7	3.95628	0.73907	1.85792	1.3731374
PRI1	4.16393	0.79098	4.16393	3.2936039
PRI2	4.20219	0.80055	4.20219	3.3640449
PRI3	4.00546	0.75137	4.00546	3.0095703
Electronic Service Quality				2.5288073

Based on the results of the E - S – Qual calculation, it is found that the value obtained was 2.5 out of a maximum scale of 5. This means that there are still many things LAZADA can further explore for the sake of customer satisfaction.

According to the dimensions measured, it is found that the dimensions of Privacy or security has the highest value which is 3.224, this signify the safety and comfort that customers feel in entering data during the process of buying and selling at LAZADA online store. Meanwhile the lowest eservqual value is of the last dimension: fulfillment, which has a maximum value of 1.3852 on a scale of 5. This proves that both customer and prospective customer of LAZADA online store find that Lazada's fulfillment of goods lack a good criterion. Among them are items sent not as scheduled and proneness to deliver goods which are inconsistent with the booking made by customer. Additionally, LAZADA is often out of stock of the goods offered which resulted in unilateral cancellation by LAZADA, further causing costumer dissatisfaction. Then an IPA matrix was depicted to determine prioritization of what customers want [7]. Priority matrix (Importance Performance Analysis) is described as figure 1.

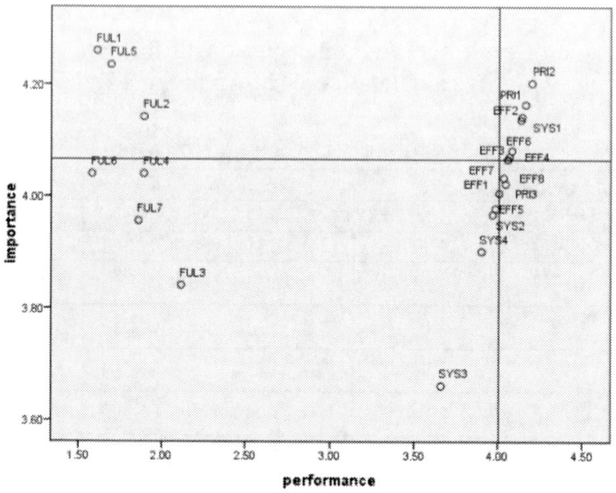

Figure 1 Importance and Performance Matrix

Based on the prioritization using importance performance analysis matrix the following results are obtained:

- There are eight factors that go into Quadrant I, namely PRI2, PRI1, EFF2, SYS1, EFF6, EFF3, and EFF4, and EFF7. This quadrant shows High Importance and High Performance, which are factors considered very important by customer and performed or provided excellently by the manufacturer or service provider. For example, in PRI2 "LAZADA do not share my personal information to another party", which means that the customers think of it as a very important factor and LAZADA provided this service excellently.
- There are three factors included in Quadrant II. The factors are EFF8, EFF1, and PRI3. Quadrant II is introduced as Low Importance High Peformance, where the factors deemed as less important by customers but are provided excellently by LAZADA. For examples is EFF8 "LAZADA Online Shopping Webite is well organized", for customers it is not too important but LAZADA provided the service splendidly. In this quadrant the producers considered to be excessive in providing services.
- There are eight factors included in Quadrant III, namely EFF5, SYS2, SYS4, SYS3, FUL3, FUL7, FUL6 and FUL4. Quadrant III is introduced as Low Importance and Low Performance, interpreted to comprise factors that is not too important for the customers and not properly provided by the service provider itself.
- Next, there are three factors included in Quadrant 4. Namely FUL1, FUL5, and FUL2. This quadrant can be considered as the most important quadrant. It is introduced as High Importance and Low Performance, which include factors deemed important by customers but has low level of performance or provision by LAZADA.

https://doi.org/10.1051/shsconf/20184901014

4. Conclusion

Based on the results of the E - S – Qual calculation, it is found that the value obtained was 2.5 out of a maximum scale of 5. This means that there are still many things LAZADA can further explore for the sake of customer satisfaction. The factors that are deemed important by customers but has low level of performance or provision by LAZADA include FUL1, FUL5, and FUL2. These factors are: Lazada's inability to send orders according to it is scheduled time, Lazada did not assure the availability of the item offered and LAZADA is often out of stock of the goods offered which resulted in unilateral cancellation by LAZADA. Proposed improvements regarding point number 2 above are the evaluation of supplier performance and goods checking on a regular basis as well as database management.

References

1. V. A. Zeithaml, A. Parasuraman, A. Malhotra, *J. of the Acad. Mark. Sci,* **30**(4), 362-375. (2002).
2. M. I. Gomez, E. W. McLaughlin, D. R. Wittink, *J of Retailing*, **80**(4), 265-278. (2004).
3. J. Santos, *Managing Service Quality, An International Journal*, **13**(3), 233-246. (2003).
4. D. Aryani, F. Rosinta, *Bisnis & Birokrasi Journal*, **17**(2). (2011).
5. A. Parasuraman, V.A. Zeithaml, A. Malhotra, *J of Service Research*,**7**(3), 213-233. (2005).
6. L. J. Cronbach, *Psychometrika*, **16**(3), 297-334. (1951).
7. J. A. Martilla, J. C. James, *The journal of Marketing*, 77-79. (1977).

URL: http://www.aitinesia.com. Diakses tanggal 24 Agustus 2016.
URL: http://www.trustedcompany.com. Diakses tanggal 23 Agustus 2016.

Design and analysis of trapezoidal bucket excavator for backhoe

Sumar Hadi[1,*], Bayuseno[1], Jamari[1], *Rachmat* Muhamad Andika[1], and *Kurnia* Chamid[1]

[1] Mechanical Engineering Dept, Diponegoro University, Semarang, Indonesia

Abstract. Excavator is the most multifunction heavy tool because it can handle other heavy tool such as sloping, dumptruck loading, and breaker. Usually, when doing excavation there is always possibility of breaking of pin in tooth adapter. The aim of this paper is designing bucket excavator using SOLIDWORK 2016 software and doing analysis with ANSYS R15.0 software using finite element analysis (FEA) method. This paper emphasize about the development that is done in the tip of excavator bucket teeth based on the calculation and ANSYS simulation with maximum shear stress result which is in the amount of 50,437 Mpa with maximum deformation level is being found at the part of excavator bucket head.

Keywords: Excavator Bucket, Trapezoidal Bucket Design, FEA approach

1. Introduction

Excavator is heavy tool consisting of arm, boom (shoulder), and bucket which is pushed by hydrolyzed strength of the handled diesel machine and trackshoe. Excavator has multifunction ability because it can handle various of other heavy tool works. As the name implies (excavation), an excavator is a piece of heavy equipment that is commonly used in mining work, construction work and work that requires lifting that can be too heavy for humans [1]. But not limited to that alone, excavators can also do construction work such as making a slope (sloping), loading dump truck (loading), breaker (breaker), etc.

The teeth of excavator is a part that gets the first contact with soil while doing excavation works in various situations [1]. Using excavator can increase some necessities in the daily life, but in its application, there is no checking works of the emergency possibilities while it is used, so that the damage of teeth bucket excavator can occur when excavation process. So, in this case explains that sometimes the damage of teeth excavator part because of some handling is not appropriate by operator. Therefore, it is necessary of doing pressure calculation at the tip of part the bucket excavator teeth using finite element method which aims to decrease the damage.

* Corresponding author: sumarhs.undip@gmail.com

2. Theory of Force Calculations

Penetration of a bucket of the material is determined by the arm crowd force (Fs) and bucket curling force (Fb). These is set by SAE J1179 standard "Hydraulic Excavator and Backhoe: Digging Forces" [2].

These rated digging forces are the forces that can be exerted at the most outside cutting point (that is the tip of part the bucket teeth). Forces calculate by applying relief hydraulic pressure to the cylinders providing the digging force.

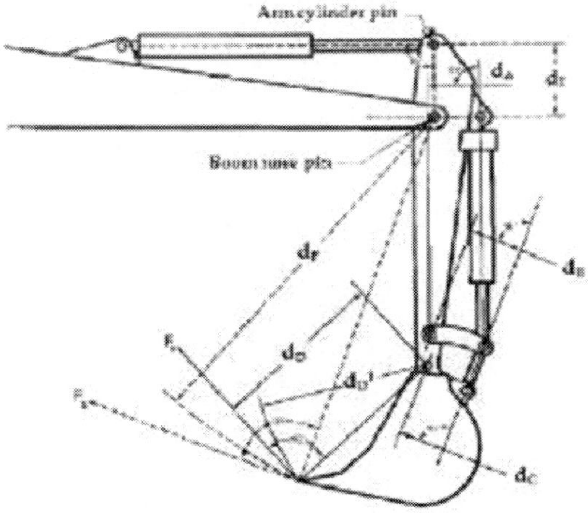

Fig.1. Determination of Digging Forces [2]

According to SAE J1179: Maximum radial tooth force through bucket cylinder (bucket retraction force) F_B is the digging force that occur by the bucket cylinder that hits the arc or radius d_D [3].

$$F_B = \frac{p \times (\pi/4) D_B{}^2}{d_D} \left(\frac{d_A \times d_C}{d_B} \right) \tag{1}$$

Arm crowd force (Fs) is force that occurs at the tip of part the bucket teeth due to the arm cylinder is the digging force generated by the arm cylinder and which touches the radius of the beam d_F.

$$F_S = \frac{p * \frac{\pi}{4} * D_A{}^2}{d_F} (d_E) \tag{2}$$

Breakout Force (L): as shown in fig. 2, forces at the tip of the teeth created by the bucket cylinder. Maximum breakout force (L) is reached when tooth force reaches maximum.

$$F1 := \frac{d}{C \times e} \times \frac{r}{L} \tag{3}$$

F 1 = Bucket cylinder force

C = Distance between bucket cylinder axis - lever pivot

D = Distance between connecting link axis - lever pivot

e = Distance between connecting link axis - bucket pivot

r = Radius bucket pivot - tooth lip

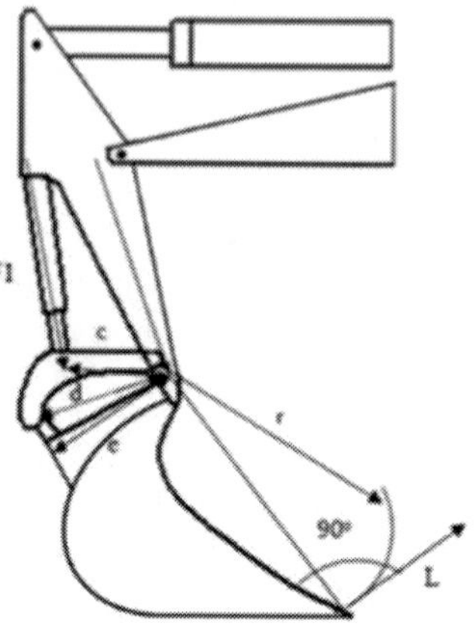

Fig.2. Determination of Breakout Force [4]

3. Methodology

This paper based on two main methods. There are geometric modelling of trapezoidal bucket and analysis methode using finite element (FEA) approach.

3.1 Material Selection

The bucket material for bucket excavator were taken as of Alloy Steel Hardox 500 standard and have the following properties as shown in Table 1.

3.2 Geometric Modelling Methode

The geometry of the trapezoidal bucket excavator has been obtained by an extrusion in SOLIDWORK 2016. First generates an trapezoidal bucket excavator then teeth-rivet, as shown in Fig. 3. Bucket and teeth-rivet are assembled together in assembly in SOLIDWORK 2016. The geometry is ready for preprocessing on next step.

Table 1. Material Properties [5]

Modulus Elasticity	210 GPa
Yield Tensile Strenght	1000 MPa

Ultimate Tensile Strenght	1250 MPa
Impact Toughness	30 J
Poisson's Ratio	0.29
Density	7850 kg/m^3
Brinell Hardness	370 - 500

Fig.3. Excavator Trapezoidal Bucket

3.3 Finite Element Analysis (FEA) Methode

The model was developed in the SOLIDWORK 2016. The element analysis method is usually used to improve the design and optimization a product of heavy tool equipment. My main discussion is on the trapezoidal bucket that comes first in contact with the soil for its contact deformation as well as the stress generation to perform various types of operations. To achieve this, trapezoidal bucket excavator have been analyzed under different boundary conditions and maximum loads.

Table 2. Mesh Statistics

Statistics	
Nodes	15641
Elements	8205

It is the imitation of the operation of a realistic process or system over time. To simulate something first, a model must be developed; this model represents the most important characteristics or behaviors / functions selected or abstract system or process. ANSYS offers various kinds of affordable technologies and services to meet the diverse and evolving design needs. In ANSYS R15.0 Stress element or solution is displayed as a different color based on the stress value to distinguish parts have a high value down to a low value. These Steps are Main Menu > Project > Model > Static Structural > Solution.

https://doi.org/10.1051/shsconf/20184902001

The reports generated by ANSYS R15.0 on solving bucket excavator are as shown in Fig. 5.

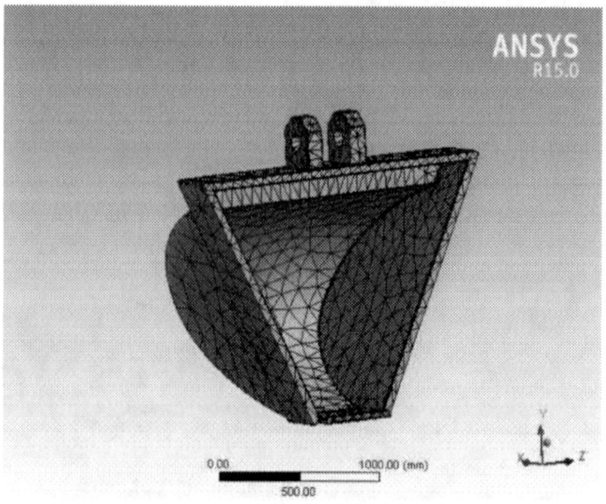

Fig.4. Meshing Trapezoidal Bucket

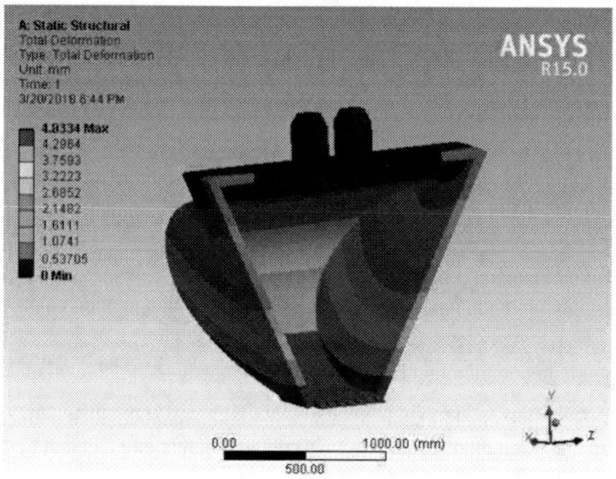

Fig.5. Total Deformation

 https://doi.org/10.1051/shsconf/20184902001

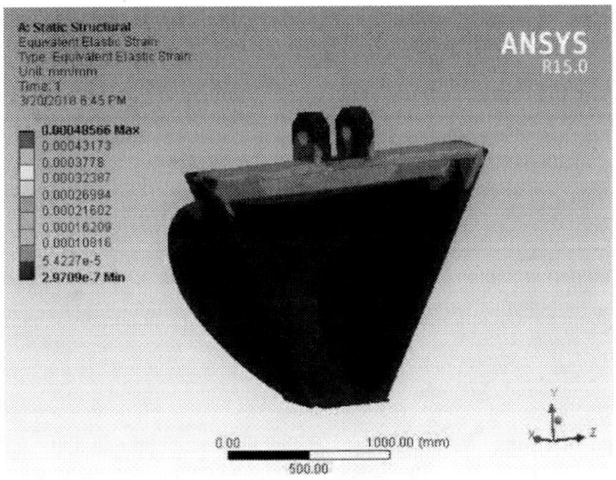

Fig.6. Von Mises Strain

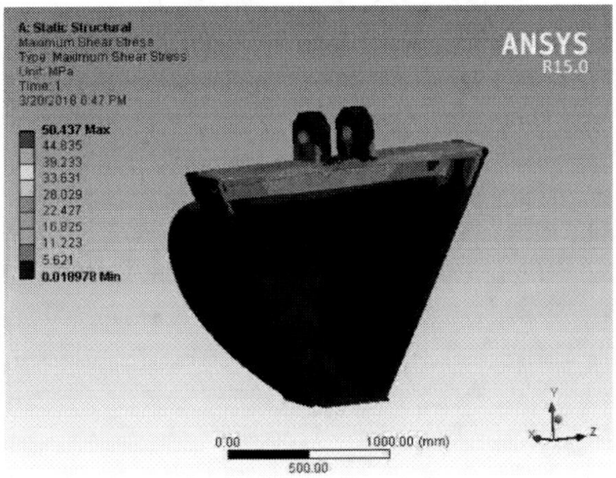

Fig.7. Maximum Shear Stress

4. Result and Discussion

The mechanical properties of steel used in linear elastic finite element analysis are also taken: Young's modulus is 210 GPa and Poisson's ratio is 0.29. This material properties assumption to be linear elastic and the region strain relation is represented by Hooke's law. Maximum Shear stress is calculated 50.437 MPa. Maximum strain using von-misses stress theory as shown in fig.6 is calculated 0.0004857 where it shows maximum strain is at the fixed part. Maximum deformation is is calculated 4.8334 mm at the tip of part the bucket teeth shown in fig.5 and thus wear of teeth occurs from tip to the entire body. It was observed that the head of tooth is the most critical point and so it could be concluded that steel with a high strength will be suitable because of the extreme loads..

5. Conclusion

This paper design and analyze a Trapezoidal bucket excavator by using ANSYS R15.0. Following conclusion can be drawn from the above analysis:

1. Maximum strain that display in a fixed position of top the bucket.
2. Maximum deformation will occur at the end of the tooth to the entire body.

As per it is recommended that buckets used for excavation purposes should be checked properly for their application based on soil. Also geometry is one of the parameters and effects of deformation during the lifetime a bucket. Given the failure of the tooth and rivet due to impact loading, it is economical to change the rivet rather than change the entire assembly of the tooth.

References

1 M. P. Tupkar and S. R. Zaveri, IJSRET, 4, 3 (2017).

2 P. P. Bhaveshkumar, IJEST. **3**, 3, 1990 – 1997 (2011).

3 SAE INTERNATIONALS, SAE J1179: *Hydraulic Excavator and Backhoe Digging Forces*, 400 Commonwealth Drive, Warrendale, PA, 1990, page 1.

4 B. Kumar and T. Alam, "*Excavator Bucket Tooth Wear Analysis*"., International Conference on Electrical, Electronics, and Optimization Techniques (ICEEOT), (2016).

5 S. Sharma and A. Sharma, "*Analysys & Opitimization of Hydraulic Excavator Bucket Teeth Using FEM*", 2017, Mechanical Engineering Department, SSTC Bhilai, Volume 03, Issue 04 (2017).

Evaluation of conventional standards of nickel electroplating process for surface finish layer of AISI 316L stainless steel using copper rod bars as cathode

Patrick Munyensanga[1,*], *Susilo* A. Widyanto[1], *Moh* N.A. Aziz[1,2], and *Paryanto*[1]

[1]Department of Mechanical Engineering, Diponegoro University,Jl. Prof. Soedarto, SH Tembalang Semarang, Indonesia.
[2]Department of Mechanical Engineering, 17 Agustus 1945 University Jl. Semolowaru no.45 Surabaya, Indonesia.

Abstract. Nickel electroplating process at UD Barito Electroplating is a modern methodology installed in an open complex environment and at controlled temperature with trust consideration of specimen composition. The evaluation of process at UD Barito has been undertaken by considering Stainless steel AISI 316L as specimen through a conventional nickel electroplating process. This paper presents the proper scientifically tool to be used at UD Barito in order to meet the time required and the surface layer quality of the electroplated product. The use of ultrasonic to replace air compressor assembly does not cause any harm to the surface finish smoothness and the quality of plated specimen given that the process is conventional. Concerning the process of 10 min compared up to 15 min of the process at UD Barito; showed the facility of ultrasonic assisted the electroplating process. This process is also a useful method in maintenance activities for surface cleaning and for product surface-layer quality enhancement through the coating process. From the data analyzed, the specimen has a hydrogen embrittlement rate range between 20 to 65% of Rockwell scale HRC52.

1 Introduction

Nowadays, alloy electroplating is widely used in the production of novel materials that require peculiar physical or chemical properties. Thin layers of iron-nickel alloys are of practical interest since these materials are greatly used in various industrial applications. Currently, small and medium industries are developing their process design, material strength, surface finish layer, as well as mechanical and chemical properties of this metal alloy. Alloy electroplating is one of the technique to improve mechanical and chemical properties of materials and has been widely used in the synthesis of Novel materials [1].

* Corresponding author: patrickmunye@gmail.com

The nature of the mechanics can increase their resistance to wear and increase the violence. The coated chemical nature can hold corrosion while physical nature attributes smooth surface side and appearance. The metal coating process can be applied by Physical Vapor Deposition (PVD), Chemical Vapor Deposition (CVD), spraying and also by electroplating process [2]. Approximately 150,000 tons of electroplated nickel annually distributed to the worldwide market. Therefore, this widespread of useful coated properties and its versatility to the materials; give nickel-electroplating features to be a modification of composition properties of the material. This modification is done through an electrolyte change in composition and the operating conditions. These properties of nickel are designed to meet the demand from the operators [3].

This paper emphasizes on the time required to coat the surface finish layer of Stainless steel 314L (Table 1) through Nickel electroplating process by comparing the method at UD Barito electroplating. Through coating process, the specimen exhibits corrosion resistance, aesthetic appeal, adhesion reduction and resistance to high temperature from Cu^{2+} electrons. Many Studies has been conducted to this material manufacturing process design with nickel electroplating with the use of Conventional Ultrasonic. This process raise temperature of electrolyte solution from 40 °C to 60 °C, [1,4,5,6].

Table 1: Chemical composition of STM 316L with314and420. (Source: Scielo)

AISI	**304**	**316L**	**420**
%C	0.0667	0.0180	1.3450
%Cr	18.1200	16.9280	13.1000
%Ni	9.5800	9.5800	5.300
%Mn	0.6020	0.6020	0.7860
%Si	0.3644	0.3644	0.4049
%Mn	0.2898	0.2898	0.1338
%P	0.0015	0.0015	0.0014

2 Metal Coating Technology

Metal coating technology is a well-known technique in metal/alloy industries to improve the quality characteristics of the material. This improvement is expected to be achieved through reparation technique in physical, mechanical and decoration features. By considering the loss of resistance to wear or failed its chemical, physical and mechanical properties, the corroded material may be maintained through electroplating depending on the nature of the material. This is rooted from the fact that mechanical characteristics can increase the resistance to wear and increase the ferocity. In addition, the chemical nature can lessen the corrosion of failed material, while physical nature presents an attribute to surface layer smooth and appearance. The coating methods were developed with the advantages of Physical Vapor Deposition (PVD).

This deposition produces a thin layer in nanometer's level of the thickness and in micrometer process in the empty air (vacuums). The PVD coaters' manufacturers are now facing the race with galvanic coatings from new condition. Although the conditions have a significant substrate of the same size to the coatings surface layer. However, the chemical and plasma of material must be cleaned and then coated at the highest possible deposition rate. These parameters provide mass production ability to the industry by using only large PVD batch systems or in-line coaters, [7].

3 Step process of Nickel electroplating for STM AISI 316L coating

The electroplating is a process of silting metal layer in the electrolyte solution that assisted by the flow of electricity. The type of solution and electricity channeled through the electrodes influences the quality of electroplating layer, [2].

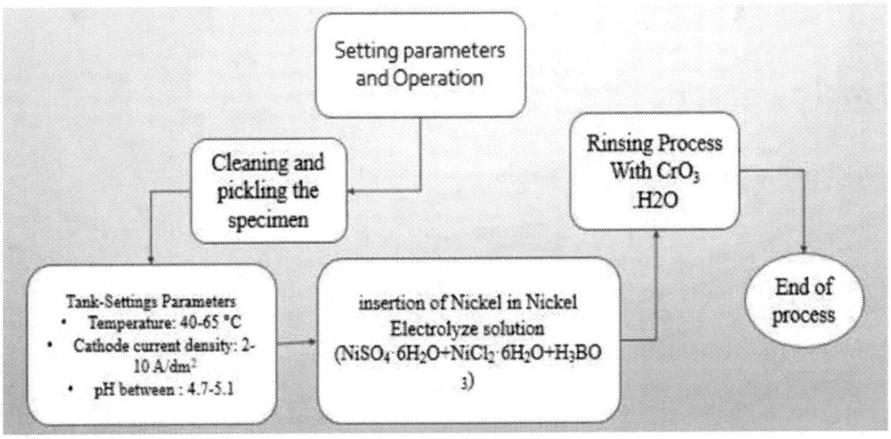

Fig. 1. Process stage of Nickel Electroplating at UD Barito electroplating

The process of electroplating (Figure 1) requires basic stage a) cleaning and pickling the surface to be electroplated b) activating coating process and maintain it after the specimen being cleaned and dried c) rinsing the STM AISI 316L after the electrolyte process. This stage includes the cleaning process with distilled water, soap and 32% of HCL for pickling.

The pickling is a combined process of cleaning the specimen STM AISI 316L. This sheets metal introduced into HCl dilute solution for 5 seconds; then blushed using distilled water (H_2O) and soap. This process aims to eliminate corrosion, dust and other contaminants on the specimen. Yuliarto in his report from the view of Mika Makinen on nickel electroplating for STM AISI 316, show the effect of the laser engine additives on machine parts with different specific features when the parts have been electroplated [8]. These additives are present in manufacturing route for the condition of the surface layer. Mika gave the definition to the results on corrosion, age, life and friction of metal surface lay that occur from different stages [2]. For example of environmental exposition stage altered to the recommended exposition from the manufacturer.

In this process (Figure 2), metal sleeve material is placed on Anode. The nickel (Ni), copper (Cu) and specimen STM 316L sheet metal to be coated have to be avail at this stage. The specimen object is placed on the Cathode and the initial electroless-plating layer of nickel offer to the surface layer a conductive tape on the specimen.

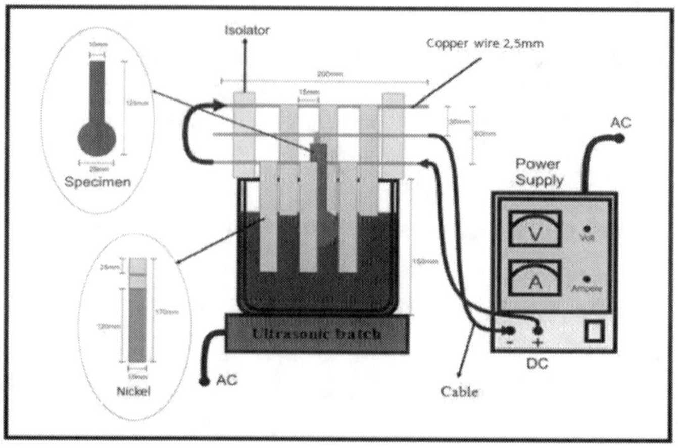

Fig. 2. Conventional nickel electroplating process diagram, (source:(2))

4 Process description

4.1. Nickel electrolyte Preparation

The pre-cleaned STM AISI 316L through pickling and brushed with soap water is now ready for nickel electroplating. The metal sleeve (copper rod bar) is placed on above of bath solution of Nickel chloride (30g of $NiCl_2$. $6H_20$), nickel sulfate (300g of $NiS0_4.6H_20$), boric acid (30g of H_3B0_3), brighteners I (15ml) and brighteners II (1 ml). Before dilution process, the bath composition dissolved in 1L of distilled water preheated by hot plate stirrer at 40^0C. Dilution process conducts simultaneously with the stirring process at 48-rpm speed rotation in 15 minutes. After the electrolyte, the solution is totally mixed and no sediment, the pH of electrolyte should be between 2 and 4.5.

The mounted stage respect the conventional process (Figure 2) as long the assembly tools chosen become intact in one unit system. This makes the electroplating system does not have the impurities or much external effect to the process. The equipment used for this process and stage facilitate easy of movement for Nickel ions Ni^{2+} and Copper ions,Cu^{2+} in 1L of the electrolyte solution.

4.2. Bath composition

The material in nickel electroplating process identified as bath composition or nickel solution. This solution composed by the chemical material of nickel sulfate (NiS04.6H20), Nickel chloride ($NiCl_2$. $6H_20$) and Boric acid (H_3B0_3). Since the process is regulated, the 300g/L of blue Nickel Sulfate solution ($NiSO_4.6H_2O$) and is intuitively provide Nickel ions during the coating process. Bath solution is also composed of 30g/L of green Nickel Chloride ($NiCl_2$. $6H_2O$) diluted. The nickel chloride improves the conductivity of nickel ions with the flow of electricity from the anode to the cathode in solution.

In the electroplating process, the PH of the solution has to be maintained at a range between 2 to 4. This range is retained by 530g/L of white Boric Acid (H_3BO_3) and 15 ml of benzene sulphonic acid. This sulphonic acid has an additional role to carry brighteners I after dilution. This brightener improves the structure of solution and results on extracted layers. For levelling agents after dilution, a 1ml of allyl sulfonic acid is used and then it fills the defect or the scratches of the substrate (specimen).

4.3. Electrolyte solution of Nickel electroplating process

The main physical supportive equipment in the solution is adjustable and metering devices. A *digital Scales Balance* to measure the mass weight of the element in nickel electrolyte solution; *Hot Plate Stirrer*, for stirring process in 10 minutes with a rotation speed of 48rpm and temperature solution at 45^0C. For distillation, the *Three Beaker glass* used for measuring brighteners I&II. Beaker serves for heating nickel electrolyte solution during the process. Beaker with 500ml for distilled HCL and Beaker with 10ml volume for liquid Brighteners. This process was conducted at variations of temperature; a Thermometer is used to measure the rise of the degree of internal heat for both electrolyte solution and dilution solution.

Stainless steel AISI 316L as Specimen to be coated and Nickel plate for maintaining the conductivity of Nickel ions Ni^{2+} during the electroplating process, this plate is a sleeve nickel metal. Copper rod bar with 2.5 mm diameter for hanger and 1.5mm for the hinge of nickel plate of 1 mm thickness. This rod bar used as connectors of power to the Cathode terminal.

Since Nickel electroplating requires a flow of electricity, a DC power supply with 30 V/5A capacity manufactured by *Jinwen inventor* is used with function purpose of transform capacity of AC voltage to the required level. In addition, it serves as a rectifier of AC current signal to DC and as a filter that ripple of AC current signal. For protecting the process to the electric circuit breakage, the Isolator or inhibitor is placed in the electric assembly. The isolator is placed between positive pole and negative pole and serve as a separator between Anode and cathode to stay intact during the electroplating process. The Red and Black connecting wires to conduct electricity and electrolyte ions (Ni^{2+} and Cu^{2+}). These two wires are binding on the two poles of the system.

This experiment design uses Ultrasonic as a heating device (heater) of electrolyte solution during the process. This facilitates fluid agitation and produces a high strength movement for sleeve ions to adhere to the substrate (specimen). It also reduces the high iron percentage in Nickel alloying process for high corrosive resistance. The experiment design replaces manual air compressor assembly and this Ultrasonic for smoothing and clearing scientifically electroplating at UD Barito electroplating factory.

4.4. Conventional Electroplating Process

The Specimen rinsed and pickled before, it is immersed into the electrolyte solution as cathode acceptor and the nickel plate is deposed into electrolyte solution attached to anode side to be dissolved and form nickel ions (N^{2+}). These ions travel through the solution and cause the formation of Copper ions Cu^{2+} on the cathode of two negative bar (Figure 3). The *STM AISI 316L* sheets metal which has been cleaned by water and soap for free of dirt and defects is immersed in the solution and hanged on copper rod bar of 2.5mm. Each of three sheets of the specimen is on one cathode rod bar; depending on the schematic of conventional nickel electroplating process diagram (Figure 2).

This process of Nickel electroplating with aid of Ultrasonic bath (Figure 3) raise the temperature from 45°C to 60°C. A thermometer is used to precise the measurement. The ultrasonic bath is connected on inventor DC power supply with a capacity of 30 V/5A plugged on PLN electricity. All required parameters have to be steady; denote the mains setup and the time for coating. The process is turned on and the control, focus on ampere meter, located on the *power supply*. The *Chronometer* is controlled also for time regulation. During the experiment, the needs for electricity has to be controlled in accordance with the requirement of the expected result.

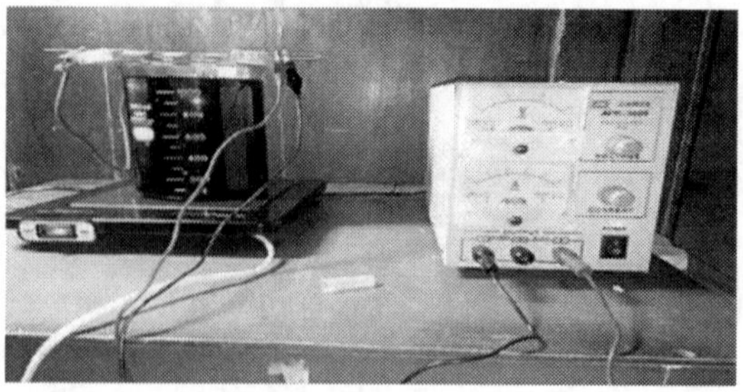

Fig. 3. Process layout of Nickel Electroplating with Ultrasonic

5 Coated Specimen Parameters Analysis

The coating process of the specimen surface finish layer is managed in 5 min. This process has been conducted at 45 °C, the initial constant temperature of the electrolyte solution. The electric current variations of 0.3A, 1A, 1.5A in this settled time with a PH level of 3.5. This phase of the test is done while the equipment coated is at free of impurities variable. This has been settled in order to meet the purpose of recovering the surface smoothness.

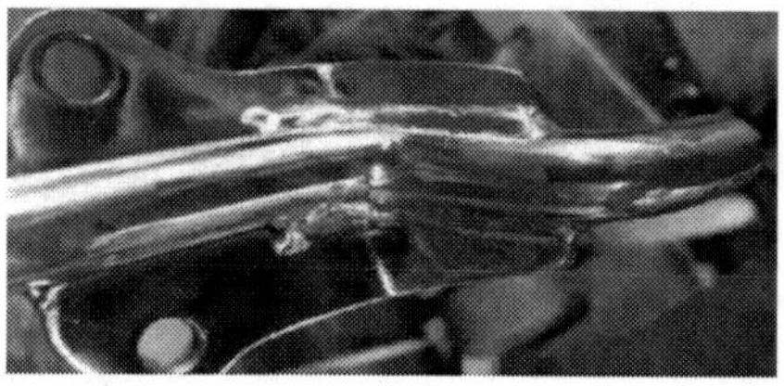

Fig. 4. STM AISI 316L specimen electroplated at UD Barito electroplating

The process used in this experiment was similar with the process implemented at UD Barito Electroplating, but the use of ultrasonic to replace Air compressor enhance agitation of electrons movement and reducing the high iron percentage in Nickel Alloy. Pickling process of the specimen is done before and after electroplating (Figure 4 by a rotating Sandpaper on an electric motor to remove all contaminants.

The Time consumed for coating was 10 min, with 2A/dm^2 plating conditions of the specimen and electric current used. A DC current rectifier with a maximum of 3000A, 20V has been used when the electricity from PLN is off. The rectifier transformed DC current of 12Volt Battery to a current ranging between 200A-250A. For the smooth surface finish, DC rectifier has been settled at 10V at the maximum operation time of 10 min. Then the Nickel plate immersed was positioned at 6 inches from one to another. During the process at UD Barito, Air is pumped into the electrolyte tank by a small compressor. After 10 min, the specimen is rinsed by an electrolyte solution of Chromium Trioxide Anhydrous (CrO_3).

This experiment has a hydrogen embrittlement rate (Table 2) range between 20 to 65% of Hardness Rockwell Scale 52 (HRC52). This ranging rate is depending on the plating parameters considered during the process. *Misumi-techcentral*, present the possibilities occurrence of the hydrogen embrittlement exist to the substrate of the electroplating process. Since this specimen used as a cathode for the electrolysis (9). The cause of this embrittlement was thought to be from agent reduction in the electroless plating solution. This effect the loss in flexibility and elasticity characteristics properties specimen.

Table 2: Hydrogen Embrittlement - Plating Embrittlement (source:(9)

Plating type	Plating condition	Hydrogen embrittlement rate (%) (HRC52)
Zinc plating		
Cyanide bath	2A/dm²、10min	68
Zincate bath	〃 〃	50
Ammonium chloride	〃 〃	4
Nickel plating		
Watt's solution bath	1A/dm²、30min	5
Wood's nickel strike bath	3A/dm²、2min	>1
Sulfamate bath		1>1
Nickel – Alloy plating	3A/dm²、20min	2
Zinc nickel alloy plating		1>1
Electroless nickel plating	90℃、1h	23
Cyanide copper plating		
Low concentration cyanide bath	4A/dm²、1min	65
Medium concentration cyanide bath	〃 〃	24
High concentration cyanide bath	〃 〃	69
Chrome plating		82

6 Conclusion

This experiment considers the chemical composition of the specimen and other variables of material. These parameters have been matched with the method used at Barito with Large batch systems, for continuous nickel electroplating process. The materials electroplated have various types of high rate deposition of impurities and corrosion on the surface layer. Pulse electroplating with Ni–W–P alloy coating has same parameters with this experiment of conventional nickel electroplating. The material pulsed has superior corrosion resistance (276.8 kΩ) and compressed surface layer without any defect manifest [9]. This result gives the specimen coated through Nickel electroplating process has a resistance capability to corrosion at a range of 276.8KΩ.

Nickel electroplating process is profitable and environmental for coating manufacturers. Since the specimen is in emergency demand. When material deteriorated or corroded, conventional electroplating method may solve time respondent of the failed device as long as the maintenance is based on servicing and surface recovering for the minimum time required in the electroplating process. The experiment conducted in 10 min proved for coated metal and with the hydrogen embrittlement range between 20 to 65 % of HRC52. The ultrasonic can cause cavitation in the solution from the agitation and it enhances the surface finish hardness and roughness of the specimen.

References

1. Poroch-Seritan, Maria Cretescu, Igor Cojocaru, Corneliu Amariei, Sonia Suciu C. Experimental design for modelling and multi-response optimization of Fe-Ni electroplating process. Chem Eng Res Des [Internet].. Available from: http://dx.doi.org/10.1016/j.cherd.2015.02.014, ;96:138–49 ,(2015)
2. Yuliarto Margen S. Elektroplating tesis sendie. (2018);
3. DiBari G a. Nickel plating. Met Finish [Internet]. Available from: https://www.nickelinstitute.org/~/media/Files/TechnicalLiterature/NPH_141015.ashx. 97(1):27, (2014)
4. Oraon B, Majumdar G, Ghosh B. Materials & Design Application of response surface method for predicting electroless nickel plating. Mater Des 27 1035–1045 [Internet]..Availablefrom:https://www.sciencedirect.com/science/article/pii/S0261306905000440;27:1035–45, (2006)

5. Chi-chang Hu, Bai Allen. Optimization of hydrogen evolving activity on nickel-phosphorus deposits using experimental strategies. Available from: https://link.springer.com/article/10.1023/A:1017525507665;5.65–72 .(2001)
6. Chi-chang Hu, Bai Allen. Composition control of electroplated nickel-phosphorus deposits. Surf Coatings Technol 137 2001 181.187T [Internet]. Available from: https://www.sciencedirect.com/science/article/pii/S0257897200011178; 2–8; (2001)
7. Panjan P, Milos I. PVD coatings as an environmentally clean alternative to electroplating and electroless processes.;119:476–87. (1999)
8. Mäkinen Mika, Jauhiainen Eeva, Matilainen Ville-Pekka, Riihimäki Jaakko, Ritvanen J, Piili Heidi SA. Preliminary Comparison of Properties between Ni-electroplated Stainless Steel Parts Fabricated with Laser Additive Manufacturing and Conventional Machining. PhysProcedia[Internet]. Availablefrom: http://dx.doi.org/10.1016/j.phpro.2015.11.048;78(August):337–46; (2015)
9. Misumi-techcentral. Surface Finishing [Internet]. misumi-techcentral. [cited 2018 Apr 12]. Available from: http://www.misumi-techcentral.com/tt/en/surface/2009/08/;(2009)
10. ZHOU, Hai hui LIAO, Zuo wei FANG, Chen xu LI H xin, FENG, Bin XU, Song CAO, Guo fei KUANG Y fei. Pulse electroplating of Ni-W-P coating and its anti-corrosion performance. Trans Nonferrous Met Soc China (English Ed [Internet]. Available from: http://dx.doi.org/10.1016/S1003-6326(18)64641-2;28(1):1–8; (2018)

Evaluation model: A performance appraisal model for certified lecturers

Dyah Ayu Puri Palupi[1,*], *Mardanung* Patmo Cahjono[1,*], and *Kristyana* Dananti[1]

[1]Universitas Kristen Surakarta, Jl. R.W. Monginsidi No. 36-38, Solo, Indonesia

Abstract. This study aims to develop lecturer performance appraisal model using web-based Integrated Evaluation Model (IEM). The object of this study is certified lecturers in Indonesia, by applying this model, the certification compensation given by the government is really used to enhance lecturer's performance. It is expected that the results of this study can be an important research that produces effective and efficient software to evaluate the lecturers' performance nationally because it involves a 360-degree assessment. We have done some activities to develop the prototype software application of Integrated Evaluation Model (IEM) for assessing the certified lecturer's performance. They are 1) preparing the questionnaire indicator for the 360 degree performance appraisal, 2) conduct the program test on the certified lecturer. In this phase, certified lecturers, superior, and students also conduct assessment on certified lecturers through IEM software, 3) conduct Focus Group Discussion (FGD) to gain evaluation and feedback on the effectiveness of the IEM application program. FGD results provide a valuable contribution to the preparation of IEM program design. And the IEM software produced in this research has accommodated all of the results of FGD activities. IEM complements the prior certified lecturer performance appraisal with 360 degrees performance appraisal.
Keywords: certified lecturers, 360 degrees performance appraisal, IEM

1 Introduction

The quality of education plays an important role in a country. Indonesian government realizes it. And the Research, Technology, and Higher Education Ministry started implementing a nation-wide program of lecturer certification with the aim to certify the lecturers in Indonesia. The ministry implemented some performance appraisal for certified lecturers.

There are some lecturers' performance appraisals in Indonesia, such as BKD, it is a Lecturer's workload including a number of jobs that a lecturer should do as an institutional task and it represents a lecturers' responsibility in three main higher education activities [1],

* Corresponding Author: dyahayupuripalupi@gmail.com, mardanung@gmail.com

SIPKD (Lecturer's Career Development Information System), it is a service for lecturers to report their activity every semester but it is under construction now (Research, Technology, and Higher Education Ministry, 2015) and SKP (Employee Job Target), it is the work plan and targets should be achieved by a civil servant. The SKP contains activities of official duties and targets should be achieved in the period of time in a measurable assessment [2]. BKD is only to evaluate the lecturers' performance based on the report and supporting data of the three main activities that the lecturers should do i.e. education, research, and community service. SIPKD assesses the performance based on the report and the supporting data of the three main activities by online. The three of the performance evaluation system have not included the assessment from students and peer review yet. Here is the existing performance appraisal (e.g. BKD/Lecturer's Workload):

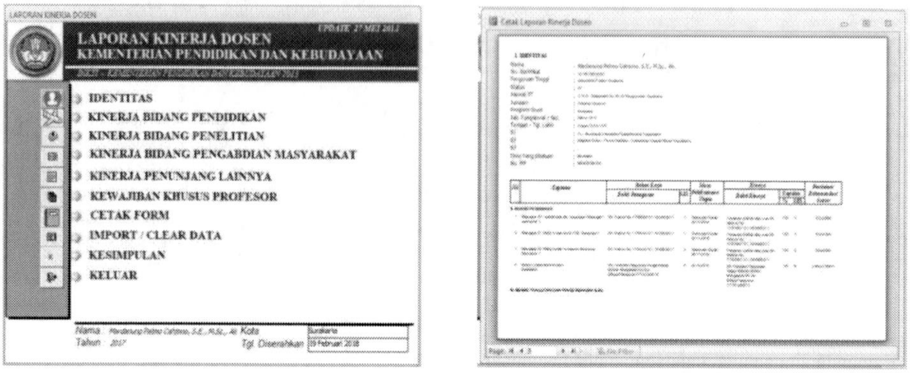

Fig. 1. BKD/ Lecturer's Workload

The fact, assessment from superior, peer, and students are really important. Because it shows the real assessment of the lecturer's character. A good lecturer is not only good at the three main activities (education, research, and community service) but also at their relationship with peer, students, and superior. So we proposed the performance appraisal model including 360 degrees evaluation, as followed:

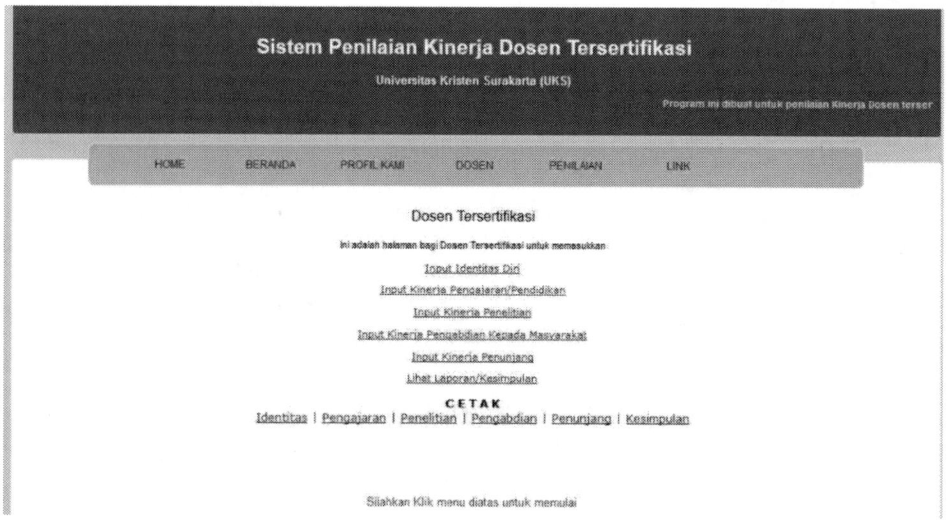

https://doi.org/10.1051/shsconf/20184902003

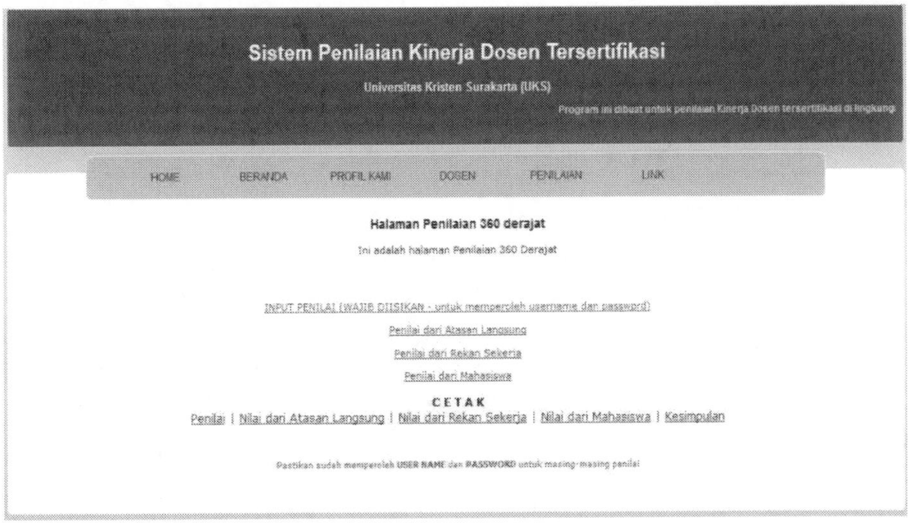

Fig. 2. Integrated Evaluation Model (IEM)

The provision of lecturer certification also still need more serious attention. According to Erlan Muliadi [3], from UIN SUKA Yogyakarta states that the certification policies that reap the pros and cons in society are still a warm conversation to talk about. Because the government policy of certification has a great influence for the direction of education in Indonesia and it is mainly related with lecturer performance after receiving certification compensation.

Certification is a proof that a lecturer has professionalism, integrity, high moral character and competitiveness. That is why the lecturers must have high integrity and moral. It is proved from their attitude to the peer, superior, and students.

2 Theoretical Review

2.1 A 360-degree Performance Appraisal System

In 360-degree appraisal system, information is obtained from several sources such as the boss, top management, assistants, co-workers, customers, dealers, advisors, and community officials. All these can be classified into internal and external parties [4]. It collects data from a variety of sources to accurately depict performance information and is a useful tool in identifying employee strengths and weaknesses. If the 360-degree appraisal is undertaken with its proper ground work, and in clear perspective it will work as a powerful tool for the performance management system [5].

2.2 Performance

Performance is the result of human behavior which is an important factor for evaluating individual work effectiveness [6]. Thus, the success or failure of an organization depends on the job performance of the individuals in that organization. Performance is an important factor for better enhancement of the organization.

Pujiwati [7] stated that performance assessment and career development is able to maintain the level of employee's commitment. Khan et.al [8] defined job performance as work performance in terms of quantity and quality expected from each employees (business dictionary).

2.3 Certified Lecturers

As stated in Law of Teachers and Lecturers Number 14 of 2005 regarding Teachers and Lecturers, lecturer is defined as a professional educator and scholar who have the main duty in transforming, developing, and disseminating knowledge, technology, and art through education, research, and community service. Professional is defined as a work or activity which is done and being an income source which requires skill and expertise which fulfill the quality standard or a particular norm and professional education. Certified lecturer is a lecturer who has earned credential from authorize source such as government. They have been given educator certification after they reported their self-description about their teaching qualification included a number of academic and professional degree [9].

3 Result and Discussion

3.1 Integrated Evaluation Model (IEM) Indicators

Based on the responses from the focused group discussion attended by 12 lecturers and 8 students, they are included languange expert, IT expert, and educational expert. The questionnaires indicators in the Integrated Evaluation Model (IEM) should be adjusted to the four competencies stated in the Law of teachers and lecturers Number 14 of 2005. The four competencies, namely pedagogic competence, professional competence, social competence, and personality competence. So we made some changes in the IEM questionnaires indicator. Before getting some suggestion from the FGD participant, we didn't include the four competencies.

Here are the questionnaires after getting some changes:

a. The questionnaire of superiors includes:

Pedagogic Competency

- Have relevant knowledge with the current position
- Accomplish the tasks given on time
- Able to communicate with the students well

Social Competency

- Consult the problems with the superiors
- Able to communicate with the community well
- Always provide an alternative solution for the problem encountered by institution
- Always provide good initiatives for the institution progress

Professional Competency

- Use the time effectively
- Arrive and leave the meeting on time
- Arrive at work on time
- Respond appropriately to feedback on job performance.

Personality Competency

- Respond feedback on critique and advice appropriately and quickly.
- Loyal to the institution

- Obey the institution's rules and regulations

b. The questionnaire of peer includes:

Pedagogic Competency

- The evaluated lecturer gives the peer suggestions as to how they can improve their teaching
- The evaluated lecturer has knowledge and understanding of learning practices

Social Competency

- The evaluated lecturer provides a helping hand for the problem encountered by colleagues
- The evaluated lecturer often shares the information related to the learning activity and institution
- The evaluated lecturer has a willing to collaborate with study program members or with faculty members
- The evaluated lecturer works to establish and maintain a positive relationship with all colleagues
- The evaluated lecturer often participates in the formal and informal activities with the colleagues

Professional Competency

- The evaluated lecturer often has a positive communication discussion about improving learning process quality with colleagues
- The evaluated lecturer is not affected by personal business when he or she works at the office

Personality Competency

- The evaluated lecturer often avoids having conflict with colleagues
- The evaluated lecturer has a good character related with the interpersonal relationship

c. The questionnaire of student includes:

Pedagogic Competency

- The evaluated lecturer masters the learning material
- The evaluated lecturer has an ability to explain the subject clearly and systematically
- The evaluated lecturer is able to motivate the students to study harder
- The evaluated lecturer is willing and able to answer the students' questions in class
- The evaluated lecturer gives the lecturer based on the syllabus

Social Competency

- The evaluated lecturer is willing to help the students outside the class
- The evaluated lecturer has an objectivity in academic assessment
- The evaluated lecturer uses the learning facilities such as LCD and whiteboard

Professional Competency

- The evaluated lecturer explains the syllabus and learning contract in the first meeting
- The evaluated lecturer gives the instruction clearly on the assigned tasks
- The evaluated lecturer assigns tasks based on the learning material
- The evaluated lecturer gives test based on the learning material
- The students can get benefit from the material

Personality Competency

- The evaluated lecturer responds appropriately on critique and advice.
- The evaluated lecturer is able to control the emotion in class well

The result of the validity and reliability test from the item above as followed:

Table 1. Validity and Reliability Output

Compe Tency	Superordinate appraisal			Peer review			Student evaluation		
	Item	Validity	Cron. Alpha	Item	Validity	Cron. Alpha	Item	Validity	Cron. Alpha
Pedagogic	3	Valid	0.845	1	_	_	6	Valid	0.793
Social	6	Valid	0.783	6	Valid	0.787	3	Valid	0.821
Professional	4	Valid	0.834	2	Valid	0.820	5	Valid	0.781
Personality	3	Valid	0.814	2	Valid	0.822	2	Valid	0.589

Source: data procressed

3.2 Integrated Evaluation Model (IEM) Development

The certified lecturer appraisal model includes:
• Administrative performance appraisal through BKD (Lecturer's workload). It consists of the assessment on education, research, community service, and supporting activities
• A 360 degrees performance appraisal, the evaluators are superior, peer, and students. The evaluators fill the questionnaires in the program. This is the novelty of the research.

The implementation of IEM on the certified lecturers showed a positive result. The students as the evaluator seemed satisfied because they could participate in evaluating their lecturers. The certified lecturers seemed satisfied too, because they were able to know their performance, not only in academic performance but also in social performance. The lecturers can get the performance appraisal result from IEM output as followed:

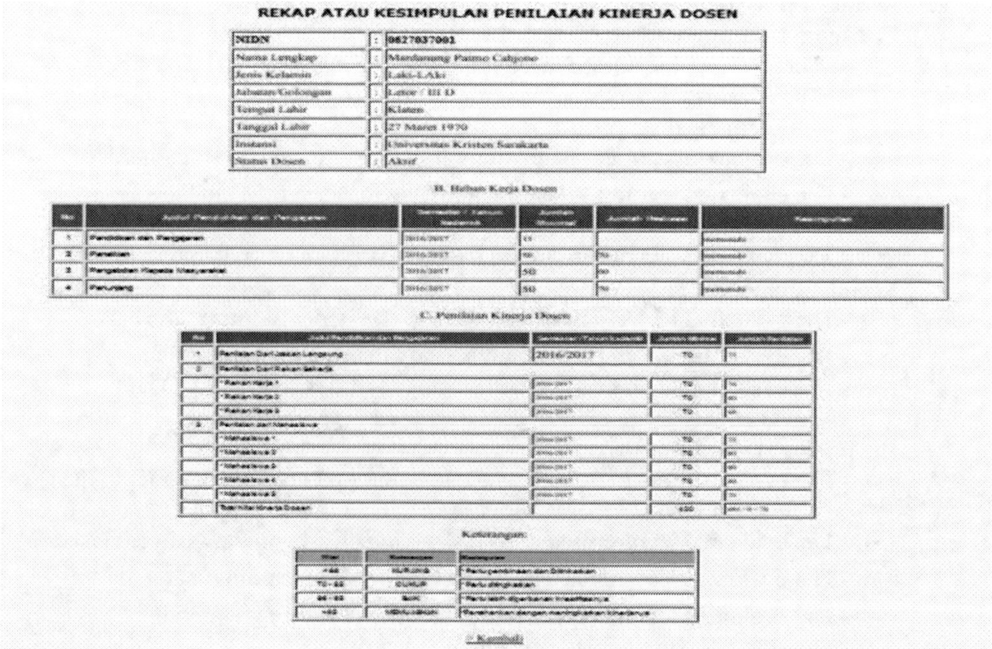

Fig. 3. IEM Output

4 Conclusion and Suggestion

4.1 Conclusion

Based on the FGD results and according to the Law of teachers and Lecturers Number 14 of 2005, the certified lecturers performance appraisal indicators are grouped into four categories, namely pedagogic category, social category, professional category, and personality category. And IEM is an effective way to evaluate the certified lecturers' performance, because it doesn't only evaluate the academic performance but also the character of the certified lecturer. And it is the novelty of this research. Here is the IEM software:

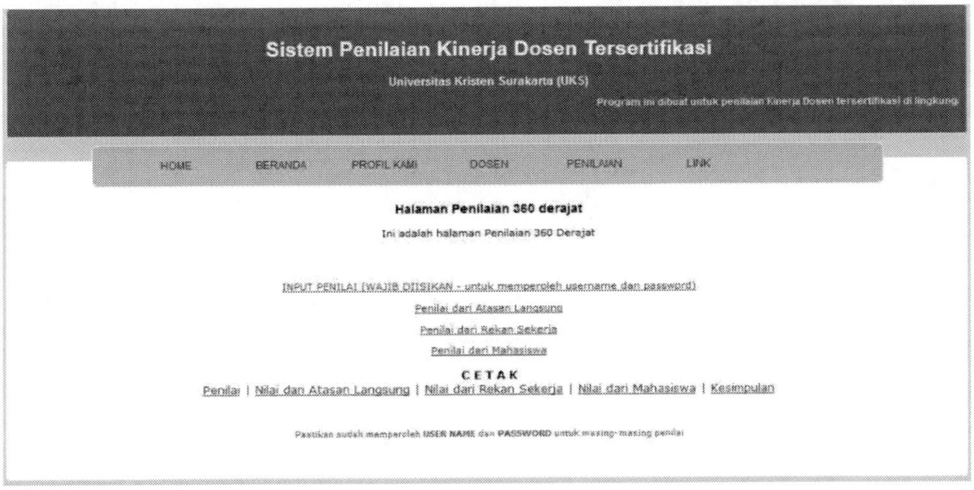

Fig. 4. IEM Software

4.2 Limitation

The limitation of this study is the respondent. They are the certified lecturers of Kopertis 6 Central Java only.

4.3 Suggestion

IEM could be implemented as the certified lecturers' performance appraisal nationally because it evaluates both academic performance and character of the certified lecturers.

Acknowledgement

I would like to express my special thanks of gratitude to Research, Technology, and Higher Education Ministry which funded us in doing this research via Applied Product Research in 2016-2017.

References

1. M. Adimihardja, Beban Tugas Dosen, Universitas Lampung (2010)
2. Regulation of Government number 46, 2011

3. http://erlanmuliadi.blogspot.com/2011/04/kebijakan-sertifikasi-guru-dan-dosen.html, accessed on Tuesday, March, 12th 2013 at. 21.07
4. MI. Rasheed, HD. Aslam, S. Yousaf, and A. Noor,. A critical analysis of performance appraisal system for teachers in public sector universities of Pakistan: A case study of the Islamia University of Bahawalpur (IUB). Af J B M, 5, pp. 3735-3744, (2011).
5. T. Ragunadhan and AH. Sequiera,. Strategic 360 Degree Performance Appraisal Model as a Synergy for Strategic Education Planning in Premier HTIs in India. Am Int J R Hum, Arts and Soc Sci. (2013)
6. J. Saetang, K. Sulumnad, P. Thampitak, and T. Sungkaew,. Factors Affecting Perceived Job Performances among Staff :A Case Study of Ban Karuna Juvenile Vocational Training Centre for Boys, J B Sci, 5, 1, pp. 33 – 45 (2010)
7. A. Pujiwati, The Influence of Performance Assessment and Career Development on Employee's Commitment. RIBER, 5, 2, 175-185 (2016)
8. MR. Khan, Ziauddin, FA. Jam, M.I. Ramay,. The Impacts of Organizational Commitment on Employee Job Performance. Eur J. of S S. 15, 3, pp. 292-298. (2010)
9. Law of Teachers and Lecturers Number 14, 2005.

Neural network method for control valve cost estimation on the EPC project bidding

Gilang Almaghribi Sarkara Putra[1,*], and *Rendra Agus* Triyono[1,*]

[1]Industrial Plant Department, PT Wijaya Karya (Persero) Tbk, Jakarta, Indonesia

Abstract. Cost estimation on the bidding phase is a crucial stage that determines the success of the Engineering, Procurement and Construction (EPC) project. If the cost offered to the client is too high then it could not compete with the other bidder, but if the cost offered are too low it can reduce profit margins and result in losses for the EPC companies. This paper describe the use of Back Propagation Neural Network method to help determine cost estimation. This method is applied specifically to determine control valve cost estimation on the bidding phase so that the retrieved costs will be accurate. When there is no technical and price quotation from vendors as well as the narrowness of the bidding processing time, this method can be an alternative choice to determine the price based on previous vendor quotation. In the future, this method could be developed and applied for other instrumentation equipment such as transmitter, switch, analyzer, control system and others to achieve total cost estimation of instrumentation equipment in EPC bidding proposal.

1 Introduction

Control valve is an instrumentation component that almost exist throughout the industrial plants mainly in the oil-gas industry and petro-chemical. On the bidding phase, control valve cost estimation count as critical item because it is included in the list of major instrumentation equipment. Control valve contributed a considerable portion of the cost in a bidding proposal. Control valve cost estimation on a less appropriate bidding phase can lead to profit and loss of the project when it was already running.

Biddings which are carried out by Engineering, Procurement and Construction (EPC) companies often have a short time limit, between two to eight weeks for accumulation to the client. Their workflow to process control valve Request for Quotation (RFQ) takes about two to four weeks depending on the workload of each control valve vendor and evaluation of their engineering team. Sometimes, control valve vendor could not meet the deadline set by EPC so that there will no price quotation from vendor. Control valve cost estimation is being inaccurate if there is no price quotation from vendor. Therefore, EPC estimators need a tools to make control valve cost estimation in a short time of bidding phase.

* Corresponding author: almaghribi@wika.co.id; rendra.agus@wika.co.id

Artificial Neural Network (ANN) is an artificial intelligence system that can change its structure to solve the problem based on the external and internal information that flow through the network. ANN can be used as a tool to make the mathematical model of cost estimation, so the accuracy of cost estimation become better. Cost estimation based ANN has been developed in Large Scale Project [1] and Transformer Main Materials [2]. Today, most of developed cost estimation software is based on ANN with various method [3-4]. The expected results of ANN system is to achieve a (one) 1 correlation coefficient value which means that estimated value is not much different with actual value, avoid overfitting cost estimation and give results of control valve cost estimation accurately.

2 Methods

2.1 Network architecture

The ANN method of control valve cost estimation presented in this paper using back propagation. Back propagation learning use gradient steepest descent method to determine the weights of neuron connections. Different learning algorithm will give different output. [5]. ANN method has been applied in cost estimation for construction project [6].

Network design consists of determining the size of network, number of neurons and layers required and type of activation function. Construction design of network architecture estimated cost of control valve with the ANN consists of three layers namely input layer, output layer and hidden layer. Input layers for this network are parameters of data as much as eleven pieces which are shown in Table 1, while the output layer is cost of control valve. Fig. 1 shows ANN architecture control valve cost estimation.

Table 1. Input Parameter Design

Parameter Design	Parameter Definition	Parameter Range
X1	Body Size	1-10 (1" – 20")
X2	Connection Rating	1-7 (150# - 2500#)
X3	Body Material	1-10 (Cast Iron – Duplex SS Forged)
X4	Trim Material	1-10 (Cast Iron – Duplex SS Forged)
X5	Actuator Type	1-9 (Single Acting Spring - Electrohydraulic)
X6	Area Classification	1-10 (Ex n – Ex ia)
X7	Valve Type	1-6 (Gate Valve – Choke Valve)
X8	Leakage Class	1-6 (I – VI)
X9	Fluid Condition	1-10 (Low – High Viscosity, Flow, Pressure, Temperature, Toxic, Erosive & Corrosive)
X10	Characteristic Valve	1-7 (Quick Opening – Hyperbolic)
X11	Average Inflation	1-7 (2010, 5.12% – 2017, 3.76%)

- Number of neuron input, hidden layer and output layer
  Number of neuron input determination is performed based on parameters design which amounted to 11, while the number of hidden layer nodes is determined based on evaluation of trial network with combinations level start from 10 to 30. The best hidden neurons combination which applied can be seen in the absence of overfitting and low value of RMSE (Root Mean Square Error) of the testing and training data.

- Activation function
  Activation function for a Back Propagation network should have several important characteristics, e.g.: continuous, differentiable and monotonous not decreasing [5].

https://doi.org/10.1051/shsconf/20184902004

Independent variable [x] in the function [f(x)] is number of multiplication between input values and weights. There is some activation function which is often used for back propagation, among others, binary sigmoid, bipolar sigmoid, and linear functions. Binary sigmoid has a range of values from 0 to 1 and is defined by following formula:

$$f_1(x) = \frac{1}{1+\exp(-x)} \tag{1}$$

Another activation function is bipolar sigmoid with range of values of 0 to 1 and is defined by following formula:

$$f_1(x) = \frac{2}{1+\exp(-x)} - 1 \tag{2}$$

And the last activation function is linear function defined by following formula:

$$f_1 = x \tag{3}$$

Network architecture of ANN is using the activation function of binary sigmoid for hidden neurons and linear activation function for output layer.

- Learning rate [α]
  Learning rate determines how efficiently and quickly the convergence of back propagation algorithm. Learning rate is a control parameter that regulates step magnitude at which the weights [W] is iterative regulated [5]. Smaller learning rate value will result slower network learning, whereas if learning rate is too large will result the oscillating system. Range of learning rate values is 0 to 1. Learning rate value usually selected based on testing data. The best learning rate value for ANN is 0.08 that obtained from the best RMSE test and there is no overfitting between range values 0.01 to 0.2.

- Momentum [μ]
  One of alternatives on updating weight value of network architecture to get the smallest RMSE value is by using momentum. The momentum formula is as follows:

$$\Delta W_{jk}(t) = \alpha\delta_k z_j + \mu\left[W_{jk}(t) - W_{jk}(t-1)\right] \tag{4}$$

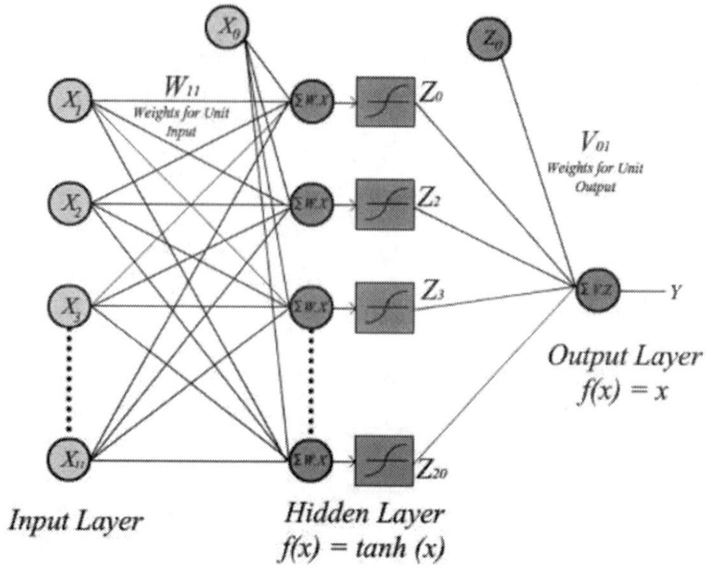

Fig. 1. Artificial Neural Network Architecture

2.2 Program algorithm

The design of feed forward network then is used for back propagation learning. This learning is done by calculating error from each layer then propagated backward to correct weights and biases in network. Program algorithm used in the training of error back propagation is gradient descent. The first step is to initialize hidden weights and output [v] weights with a random value. Then enter input value to each layer of the neuron unit by following formula:

$$Y_j = f\left(v_{0j} + \sum_{i=1}^{n} x_i v_{ij}\right) \tag{5}$$

Calculation above is used in hidden neuron and output value, from these results then do the algorithm of back propagation of its output value. First step is calculate the error value from target minus output. Then calculate the output weights correction by equation (4) above where [δk] is output error, then these weights correction we calculate as the equation (5) to obtain the hidden output error. Hidden output error value is calculated again with equation (4) so that obtained hidden neurons correction. Then update the weights by summing old weights with new weights to the hidden neurons weights and output weights. The last step is to calculate the RMSE if it's appropriate, if it doesn't repeat the first step to get the expected value of the RMSE.

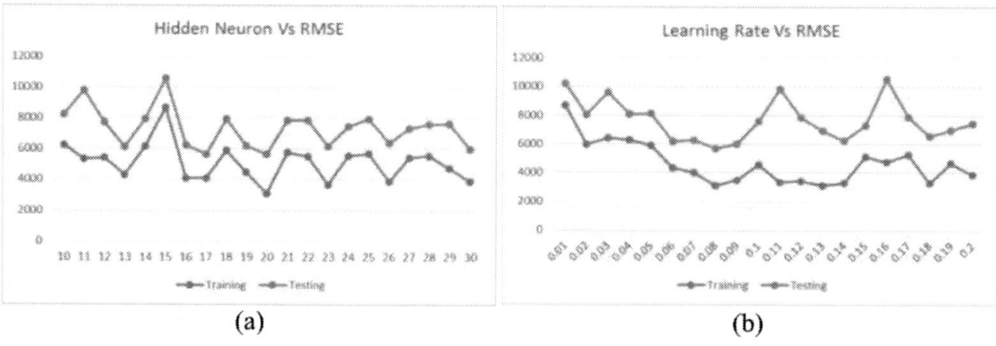

 https://doi.org/10.1051/shsconf/20184902004

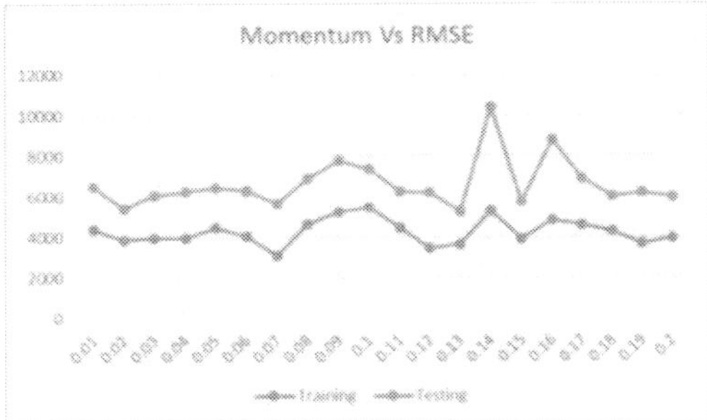

(c)

Fig. 2. a. Graphic of Relationship between Hidden Neuron and RMSE
b. Graphic of Relationship between Learning Rate and RMSE
c. Graphic of Relationship between Momentum and RMSE

3 Result discussion

3.1 Network parameter test

Network parameters test is performed by searching the best learning of combination execution between hidden layer, learning rate and momentum. Number of hidden layer neurons determine the ability of network to capture data patterns and produce nonlinear mapping between input and output. ANN software uses one hidden layer. Number of hidden layer neurons is determined through testing which is done on the network by varying number of hidden layer neurons to know network stability. Magnitude of RMSE network with numerous combination of hidden neurons is shown in Fig. 2.a, where the smallest RMSE value of hidden neurons is 20. The best learning rate value based on the results of RMSE experiments are between values range of 0.01 to 0.2. Based on Fig. 2.b and Fig. 2.c, it is shown that the best learning rate that is 0.08 and the best momentum is 0.07.

3.2 Cost estimation performance

Control valve cost estimation learning is performed with back propagation learning with number of nodes hidden neurons parameter are 20, learning rate 0.08 and momentum 0.07. Below Fig. 3 described RMSE value from learning process.

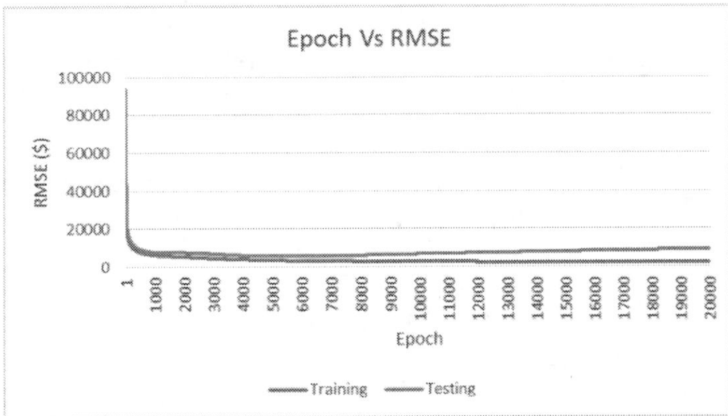

Fig. 3. Graphic of Relationship between RMSE and Epoch for Testing and Training Data

 https://doi.org/10.1051/shsconf/20184902004

Fig. 3 shows the changing of training and testing RMSE value to epoch is decreasing and there is no overfitting, where training and testing value has a descent tendency. From Fig. 3 above also can be seen that the ANN system can generalize the control valve cost estimation system, so it can be used for testing data outside of the learning system.

Network validation is done by making a regression plot by calculating the correlation coefficient (R) value. Regression plot shows the relationship between network output value and target of observation data. Blue line shows the best match value of regression line between target and output. Correlation coefficient (R) value is a value which states linear relationship between target and output that has a value of 0 to 1. If the value approaches 1, then the linear relationship between target and output will be bigger. Fig. 4.a and Fig. 4.b shows the regression plot for training and testing data.

Fig. 4.a and Fig. 4.b below shows the results of training data correlation coefficient by 0.99998 while correlation coefficient of testing data is 0.94352. This indicate that the result of estimated value and actual value is same for training data, whereas for testing data ,the result of estimated value is approaching true value.

Testing data of 28 control valve shows 91.81% average in accuracy. ANN software has a standard deviation of 5.65% and the average error to 8.19%. This result is good enough to be applied in the control valve cost estimation on the bidding phase.

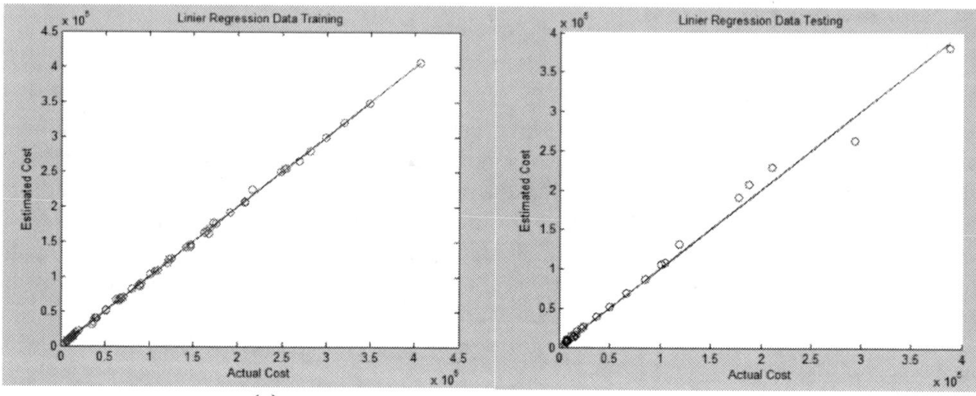

(a) (b)

Fig. 4. a. Graphic of Relationship between Hidden Neuron and RMSE
b. Graphic of Relationship between Learning Rate and RMSE

4 Conclusion

Artificial neural network method with back propagation learning algorithm can be implemented to determine the advisability of control valve cost estimation in bidding phase. Back propagation algorithm has been trained well and provide the accurate output for testing data, so offered cost estimation that is given to the client has a small error. The characteristic of back propagation which is generalization, make this system can be used for early estimation on control valve cost determination. Correlation coefficient result of data testing is 0.94352, it means that the estimation value resemble the actual value. The expected output target is accurate in testing, as it occurs adjustment of weight value and better bias on learning process.

References

1. Hui Zhao, Weijie Huang, Shengping Wang, Tao Liu. *Intelligent Cost Estimation Method for Large scale Projects Based on PCA Optimized RBF Neural Network*. 8th International Symposium on Computational Intelligence and Design. IEEE. (2015)
2. H.D. Mehta, R.M. Patel, P.H. Trivedi. *Cost Estimation of Transformer Main Materials using Artificial Neural Networks*. Nirma University International Conference on Engineering. IEEE. (2012)
3. D.Manikavelan, R. Ponnusamy. *Software Cost Estimation by Analogy Using Feed Forward Neural Network*. Information Communication and Embedded Systems (ICICES). IEEE. (2014)
4. Lixue Xia, Rong Luo, Bin Zhao, Yu Wang, Huazhong Yang. *An Accurate and Low cost PM 2.5 Estimation Method Based on Artificial Neural Network*. Design Automation Conference, 20th Asia and South Pacific. IEEE. (2015)
5. L. Fausett. *Fundamentals of Neural Networks : Architecture, Algorithms and Applications*. Prentice Hall, Inc : New Jersey. (1993)
6. Zeyan Du, Binyong Li. *Construction Project Cost Estimation Based on Improved BP Neural Network*. International Conference on Smart Grid and Electrical Automation. IEEE. (2017)

Enhancement method for assignments of staff in multiple events

Ahmad Diponegoro[1,*]

[1]University of Muhammadiyah Prof. DR. HAMKA, Graduate Program of Management, Jakarta 12790, Indonesia

Abstract. A problem of staff assignment in an organization with multiple assignments of scheduled events is addressed here. The objective is to propose an efficient and practical approach for enhancing a given policy of staff assignments to attain a more efficient operation. To address the problem, a multi-period assignment model formulation is employed here. Using this approach, an efficient heuristic method is devised to obtain enhancements of any given solution policy. A numerical example is provided for clarity in implementing the proposed method. This research is expected to provide a meaningful contribution to managing staff assignments in various types of organizations.

1 Introduction

Staff assignments play an important role in managing resources in many organizations, especially the human resource, in order to stay efficient while being pressed for achieving their operational requirement. When an assignment of staffs occurs in multiple periods, the problem usually becomes complicated by the need to match one or more requirements simultaneously, such as budget limit, skill criteria, timing, or career demand of the employee. Staff assignments in multiple periods can be categorized as a multi-period assignment problem. Obtaining an optimal solution using exact mathematical programming for a large instance of the problem in terms or the number of personnel, positions, and events, are computationally prohibitive.

Multi-period assignment problems arise in many instances, which include organizational training programs [1], doctor assignment ([2] and [3]), in scheduling law enforcement agency to elections [4], and in assigning airline customer service agent [5]. Such problems have attracted many researchers for years in which it is formulated as a multi-dimensional assignment problem [6]. It can also be formulated as a multiperiod assignment problem ([7] and [2]).

Optimal solution methods to such a problem have been sought by several researchers, such as provided by [8], [9], [10], and [11]. Since the multi-period assignment is categorized as an NP-hard complexity ([6] and [12]), exact mathematical approaches can be computationally intractable. Efficient heuristics are then sought by many as an alternative approach for tackling large instances of the problems. Most of the heuristics are

*Corresponding author : adipone@uhamka.ac.id

usually developed for a specific subclass of problems. This research enriches the literature by proposing an alternative heuristic approach for a specific staff assignments problem that arises in educational organizations that hold multiple training events, which is described in the next section.

2 Description of the problem

The assignment problem addressed here pertains to a situation in a typical training board of a government institution where several training events are planned. Employees of the organization must be assigned to choices of staff positions of training events, or referred to as events for compactness, scheduled in an annual planning horizon. Placement of an employee into a staff position of events shall accord to the requirements of the corresponding staff position, such as the employee's rank, specialization, fee, or other criteria. Hence, a weighting preference or 'cost' is applied in each staff position to provide some sort of suitability measure of assigning a person into a staff position. The less suitable a person for a staff position the higher the cost of assigning her to the position. Conversely, the more suitable the lower the cost. It is assumed that a prescribed number of assignments in the planning horizon is determined for every employee. The objective of solving an assignment problem addressed here is, therefore, to minimize the total cost of placing a set of assignments of staff in the planning horizon.

2.1 Problem formulation

To describe the mathematical problem formulation, the following symbols are employed.

K = number of events in the planning horizon;
L_k = number of staff position in event k;
N = total number of person (employee) to be assigned;
x_{ijk} = decision variable of person i assigned on j-th staff position of an event k. The value $x_{ijk} = 1$ if the corresponding person is assigned or, otherwise, $x_{ijk} = 0$ if none is assigned;
a_{ijk} = 'cost' of assigning person i into position j of k-th event;
b_i = predetermined total number of assignment for person i into all scheduled events.

Since the objective of the problem addressed here is to minimize the total cost incured by assigning the employees into staf position, the following objective function is defined as.

$$Z = \sum_{k=1}^{K} \sum_{j=1}^{L_k} \sum_{i=1}^{N} a_{ijk} x_{ijk} \tag{1}$$

Note that (a) each person can only be placed in exactly one staff position in an event; and (b) in each staff position of an event there must be assigned exactly one person for the position. Hence, the assignment problem is defined as follows. Find x_{ijk} for $i = 1, \ldots, N$; $j = 1, \ldots, L_k$; and $k = 1, \ldots, K$ so as to minimize Z subject to

$$\sum_{j=1}^{L_k} x_{ijk} \leq 1 \text{ for } x_{ijk} \leq 1; \text{ for } k = 1, \ldots, K; \text{ and for } i = 1, \ldots, N, \tag{2}$$

$$\sum_{i=1}^{N} x_{ijk} = 1 \text{ for } j = 1, \ldots, L_k \text{ and for } k = 1, \ldots, K, \tag{3}$$

$$\sum_{k=1}^{K} \sum_{j=1}^{L_k} x_{ijk} = b_i \text{ for } i = 1, \ldots, N, \tag{4}$$

$$x_{ijk} = \{0, 1\} \ \forall\, i, j, k \tag{5}$$

where

$$x_{ijk} = \begin{cases} 1 & \text{if person } i \text{ is assigned to staff } j \text{ position at event } k \\ 0 & \text{if none is assigned to staff } j \text{ position at event } k \end{cases} \tag{6}$$

2.2 Constraints and computational complexity

The first constraint (2) assures a person cannot be assigned to more than one staff position in each event. The second constraint (3) assures that each staff position in an event must be assigned a person to it, i.e., a blank staff position is not allowed. The constraints (4) is the pre-determined total number of assignment of each person in the planning horizon. In this formulation, it is assumed the constraints (4) should be binding. The last constraint (5) assures the decision variables take binary value.

The problem presented in (1) – (5) above can be categorized as a special case to a Multi-Period Assignment Problem. Specifically, it is of the three-dimensional assignment problem, in which the problem is of an NP-Hard complexity. A solution to the problem using an exact mathematical approach, such as Branch and Bound can be computationally intractable for a large size of instances. A more efficient approximation method approach is sought in this research in order to obtain an acceptable solution within a reasonable time. The following section will describe the method.

3 Methodology for assignments and computational application

The solution method sought here takes two general steps: (i) To obtain an initial, feasible solution to a given problem; and, (ii) to perform an enhancement step to obtain a better objective value. The solution method for a feasible solution in the first step is assumed to be obtained from the work by [1]. Hence, it is focused here to develop a method for the second step, i.e., the enhancement solution, which will be described next.

3.1 Enhancement method

The enhancement method proposed here consists of two phases of iterations, in which incremental improvements of assignments are sought in each phase. In the first phase of iterations, improvements of assignments are sought within each event. Here, every pair of two assignments within the same event is investigated for possible exchange in assignment positions that will result in incremental improvement of the objective function without violating feasibility to the given constraints. In the second phase, improvements of assignments are sought between every pair of two assignments from two different events. The mechanism to obtain incremental improvement is the same as that in the first phase except that it is done between two different events. The enhancement procedure is given as follows:

1. *Phase I. For* each event k do the following :
   For each pair of two assigned persons within the event k, say, person c assigned to position e and person d assigned to position f, *do* :
   a. *If* $a_{cek} + a_{dfk} > a_{dek} + a_{cfk}$ then switch the assignment positions between the two persons, i.e., the person c is assigned to position f and the person d is assigned to position e.
   b. *Else*, keep the current assignment of the corresponding pair.
2. *Phase II. For* each pair of two different events, say, events k and l, do the following :
   For each pair of two assigned persons, say person c assigned to position e in the event k and person d assigned to position f in the event l, do :
   a. *If* $a_{cek} + a_{dfl} > a_{dek} + a_{cfl}$ then switch the assignment positions between the two persons, i.e., the person c is assigned to position f in the event l and the person d is assigned to position e in the event k.
   b. *Else*, keep the current assignment of the corresponding pair.
3. *Stop*. □

It can be observed in the foregoing procedure of the first phase there are L_k by N iterations. In the second phase, there are K by L_k by N iterations at most. Taking the largest, therefore, the computational complexity of the proposed procedure is of the cubic order polynomial complexity. The following section is provided to illustrate the application of the proposed approach to a problem instance.

3.2 Computational application

Consider a hypothetical situation where a training division is planning four events of training in the next year. There are four persons to be assigned to three staff positions in every event. Each person must be assigned three times in the given planning horizon. Hence, we have $N = 4$, L_k's = 3, $K = 4$, and b_i's = 3. A linear scale of weights 1, 2, 3, 4 and 5 is utilized to represent the costs indicating one's suitability for being assigned to a certain staff position. The value of 1 indicates the 'most suitable' and 5 indicates the 'least suitable.' Suppose the costs of assigning employees to various staff positions, i.e. the a_{ijk}'s, are shown in the following matrix (Fig. 1).

Event k	1			2			3			4		
Staff j	1	2	3	1	2	3	1	2	3	1	2	3
Person i 1	4	5	1	4	4	5	3	2	4	3	5	4
2	5	4	3	2	3	1	4	1	2	4	1	1
3	5	3	1	2	3	1	4	3	5	2	1	1
4	2	2	5	4	2	2	2	3	1	5	5	1

Fig 1. Cost matrix (a_{ijk}) for staff assignment.

It is assumed that an initial assignment solution is given by applying the method given by [1]. The initial solution is given as $x_{311} = x_{221} = x_{131} = x_{112} = x_{422} = x_{232} = x_{113} = x_{323} = x_{433} = x_{414} = x_{324} = x_{234} = 1$ and the remaining x_{ijk}'s are zeros. The initial solution is depicted by the following matrix in Fig 2. The corresponding objective value (1) of the initial solution is $Z = 31$. Our proposed method will be implemented next to seek an enhancement of the given foregoing initial solution.

Event k	1			2			3			4		
Staff j	1	2	3	1	2	3	1	2	3	1	2	3
Person i 1	0	0	1	1	0	0	1	0	0	0	0	0
2	0	1	0	0	0	1	0	0	0	0	0	1
3	1	0	0	0	0	0	0	1	0	0	1	0
4	0	0	0	0	1	0	0	0	1	1	0	0

Fig. 2. Initial solution of the staff assignment.

Upon executing *Phase I* (within-event) of the enhancement procedure, it is found two pairs of staff assignments where switching position in each pair of the assignments yield better results. Specifically,

a) In Event 1: Since $a_{311} + a_{131} = 6 > a_{111} + a_{331} = 5$ then the position of Person 3 in Staff 1 and the position of Person 1 in Staff 3 are switched each other so that the position Person 3 is now assigned to Staff 3 and the position of Person 1 is now assigned to Staff 1. Thus, $x_{311} = x_{221} = 0$ and $x_{111} = x_{331} = 1$, yielding incremental reduction of $(a_{311} + a_{131}) - (a_{111} + a_{331}) = 1$ of the objective value.
b) In Event 4: Since $a_{414} + a_{234} = 6 > a_{214} + a_{434} = 5$, the position of Person 4 in Staff 1 and the position of Person 2 in Staff 3 are switched each other. This yield an incremental reduction of $(a_{414} + a_{234}) - (a_{214} + a_{434}) = 1$ of the objective value.

Event k	1			2			3			4		
Staff j	1	2	3	1	2	3	1	2	3	1	2	3
Person i 1	1	0	0	1	0	0	1	0	0	0	0	0
2	0	1	0	0	0	1	0	0	0	1	0	0
3	0	0	1	0	0	0	0	1	0	0	1	0
4	0	0	0	0	1	0	0	0	1	0	0	1

Fig. 3. Enhancement solution from Phase I.

Fig. 3 shows the resulting enhancement of Phase I with the corresponding objective value (1) is $Z = 29$. Upon executing the *Phase II* (between-events) on the foregoing result, there is a pair of staff assignment that can be switched to obtain an incremental enhancement. That is, the position of Person 2 in Staff 2 of the Event 1 is switched with the position of Person 4 in Staff 3 of the Event 3, which yield an incremental reduction of $(a_{221} + a_{433}) - (a_{421} + a_{233}) = 1$ of the objective value. Hence, the final solution is as $x_{111} = x_{421} = x_{231} = x_{112} = x_{422} = x_{232} = x_{113} = x_{323} = x_{233} = x_{214} = x_{324} = x_{434} = 1$ and the remaining x_{ijk} 's are zeros. The corresponding objective value (1) of the final solution is $Z = 28$, which is approximately 90% of the initial objective value (improvement).

4 Concluding remarks

This research proposed an enhancement methodology for assigning staff to multiple scheduled events. The method is intended to seek an improvement of a given initial staff assignment to obtain an assignment with better objective values. It is shown that the iterative approach given here is practical and efficient. The proposed method is expected to provide a useful tool for managerial decision making that deals with large instances of staff assignment. The problem addressed here has assumed the prescribed number of assignment of each employee is translated into binding constraints, as shown in (4). To provide a more adaptable model in representing many real situations, however, the assumption of binding formulation shall be relaxed. Hence, further research can be directed for tackling problems with non-binding constraints for prescribing a number of assignment.

References

1. A. Diponegoro and F. Rukman, "Modeling of a scheduling method for organizing training assignments," in *Proc. of the 2nd Annual International Seminar on Transformative Education and Educational Leadership (AISTEEL 2017)*, Medan, Indonesia, pp. 184-187 (2017), http://dx.doi.org/10.2991/aisteel-17.2017.39.
2. L. S. Franz and J. L. Miller, "Scheduling medical residents to rotations: solving the large-scale multiperiod staff assignment problem," *Oper. Res.*, **41** (2), pp. 269-279 (1993), https://doi.org/10.1287/opre.41.2.269.
3. J. Guo, D. R. Morrison and S. H. Jacobson, "Complexity results for the basic residency scheduling problem," *J. Sched.*, **17** (3), pp. 211–223 (2014), https://doi.org/10.1007/s10951-013-0362-9.
4. B. Nag, "A MIP model for scheduling India's General elections and police movement," *OPSEARCH*, **51** (4), pp. 562–576 (2014), https://doi.org/10.1007/s12597-013-0160-3.
5. Y. -H. Kuo, J. M. Leung and C. A. Yano, "Scheduling of multi-skilled staff across multiple locations," *Prod. Oper. Manag.*, **23** (4), pp. 626–644 (2014), https://doi.org/10.1111/poms.12184.

6. K. C. Gilbert and R. B. Hofstra, "Multidimensional assignment problems," *Decision Sci.,* **19** (2), pp. 306-321 (1988), https://doi.org/10.1111/j.1540-5915.1988.tb00269.x.
7. L. S. Franz, H. M. Baker, G. K. Leong and T. R. Rakes, "A mathematical model for scheduling and staffing multiclinic health regions," *Eur. J. Oper. Res.,* **41** (3) pp. 277-289 (1989), https://doi.org/10.1016/0377-2217(89)90249-X.
8. A. J. Robertson, "A set of greedy randomized adaptive local search procedure (GRASP) implementations for the multidimensional assignment problem," *Comput. Optim. Appl.,* **19** (2), pp. 145–164 (2001),https://doi.org/10.1023/A:1011285402433.
9. E. L. Pasiliao, "Local eighborhoods for the multidimensional assignment problem," in *Dynamics of Information Systems: Theory and Applications*, New York, Springer New York, pp. 353-371 (2010).
10. M. Larsen, "Branch and bound solution of the multidimensional assignment problem formulation of data association," *Optim. Method. Softw.,* **27** (6), pp. 1101-1126, (2012), https://doi.org/10.1080/10556788.2011.648931.
11. C. Vogiatzis, E. L. Pasiliao and P. M. Pardalos, "Graph partitions for the multidimensional assignment problem," *Comput. Optim. Appl.,* **58** (1), pp. 205–224 (2014),https://doi.org/10.1007/s10589-013-9619-7.
12. M. L. Fisher, R. Jaikumar and L. N. Van Wassenhove, "A multiplier adjustment method for the generalized assignment problem," *Manage. Sci.,* **32** (9), pp. 1095-1103 (1986), https://doi.org/10.1287/mnsc.32.9.1095.

Development of accounting information system and accounting standards for small and medium enterprises (SME)

Endang Satyawati[1,*], *Mujiyono*[1]

[1]The Faculty of Economics, Universitas Kristen Surakarta, Jl.Monginsidi No.36-38, Surakarta, Jawa Tengah, Indonesia

Abstract. At this time, small and medium enterprises play an important role in improving the economy of Indonesia, in terms of business number, job creation, and growth of national Economy. At this time many small, and medium enterprises are still in the manual recording process of financial transactions and financial reporting. The objective of this study is to develop an accounting information system method with Rapid Application Development (RAD) for SME as to answer whether there is an increased quality of information system in SME. The results of this research are as follows: 1) The small business enterprises are dominated by human resources. 2). Many of the owners of the small business enterprises do not understand the accounting and financial report system well. They operate with manual and conservative descriptions with a simple book noted. They have not known or have not been accustomed to computerized accounting system as well as have not known whether or not it is useful to support their business continuity. 3) Personal expenses and operational expenses paid by using the same sources.
Keywords: Accounting Information System, Rapid Application Development, Accounting Standards, SME

1. Introduction

The Indonesian economy is experiencing erratic fluctuations, which makes the small and medium enterprises become a good vehicle for the creation of productive employment as the production process of the small and medium scale industries are generally labor-intensive. Until now, the micro, small, and medium scale enterprises play an important role in improving the economy of Indonesia in terms of businesses number, job creation, and national economic growth. The small and medium enterprises is one of the keys for this nation out of the crisis, because in majority the businesses related to the real sectors are the micro, small, and medium enterprises, and most people believe that one of the keys for this nation to out of the crisis is by driving the real sectors.

* Corresponding email: endangsatyawati@uks.ac.id

1.1 Accounting Information System

At this many small, and medium enterprises are still in the manual recording process of financial transactions and financial reporting. It is not effective and efficient in business operations. The manual system often goes wrong, especially if the transaction data are abundant, which may cause the financial information presented to be not accurate, relevant, and timely.

The development of technology, especially technology information, which instrumental in all fields is very fast. In the field of business, information technology is very favorable for the survival of companies. The companies rely on information systems to maintain their competitiveness. The development of information technology is expected to improve the enterprises' progress in their business. The small and large scale enterprises need information system immensely as they need a system, which connect between the units there. The system is very important to support the activities of the companies so as to increase their productivity.

The application of information technology in the enterprises has a very large role, especially in the management and processing of data. The information generated by a computer-based information system, will meet the needs of the users. A good information system is a system of mutual support sub-systems, and is equipped with sophisticated computers.

The use of computer technology in accounting information systems or so-called Electronic Data Processing (EDP) will help improve the efficiency and effectiveness of the system itself. However, this condition can be met provided that specially made software or special applications must meet the accurate and true standards of internal control procedures. Computer technology is not only used by large companies, but with the rapid advancement in technology and the need for fast and precise information, the small and medium scale companies also need to improve the effectiveness of the work. The small, and medium enterprises in improving their business require a support for the daily operation data processing.

The rapid technological development and the need for accurate, relevant, and timely information are expected by the internal and external users of information of the companies. However, in reality there are still many SMEs which use manual systems in their data processing operation. They require a support of computer technology. The use of manual systems with an emphasis on human power can lead to several problems as follows:

1. The data processing takes long time
2. The calculation contains errors or inaccuracy
3. The results of data processing are not timely reported
4. The data security is improperly guaranteed

1.2 Accounting Standards

Outline of systems relating to the current major accounting standards In Indonesia, Articles 66 to 69 of the Company Act stipulate the contents and procedures of financial statements that should be included in the annual report. Companies are required to prepare their financial statements in accordance with the accounting standards established by the Indonesian Accountants Association, Ikatan Akuntan Indonesia (IAI).

As for accounting standards for SME, Standar Akuntansi Keuangan untuk Entitas Akuntabilitas Publik (SAK-ETAP) established the accounting standards for Non Publicly Accountable Entities (ETAP) in 2009. The standards have been applied from the fiscal year starting 1 January 2011. IFRS has been applied since 2012, but IFRS for SME is not completely adopted as the accounting standards in Indonesia. However, in the process of developing the Tier2 SAK ETAP standards, the Indonesian Financial Accounting Standards Board, an organisation inside the Institute of Certified Public Accountants (DSAK IAI), has referred to IFRS for SME and is incorporating important points in the country's accounting standards. Other related movements include IAI's announcement of the Exposure Draft of Accounting Standards for Small and Medium Entities (ED SAK UKM) in December 2008, which incorporated some amended concepts of IFRS for SME into SAK ETAP. 2017 [1]

2. Theory and Development of Hypotheses

2.1. Contributions Small, and Medium Scale Enterprises (SMEs) to Economy

Law No. 2008 regards SMEs as a legal basis and support for the empowerment of SMEs. The statistic shows that the number of units of SMEs approaches 99.98% of the total business units in Indonesia. The contribution of SMEs to economy is clearly shown by its dominance as much as 99.04% to provide jobs. Similarly, its contribution to the Gross Domestic Product (GDP) of non-oil is quite convincing that is 63.11%. The role of SMEs in Indonesia as well as in other countries is the backbone of the national economy.

Table 1: Contribution of Small and Medium Scale Enterprises (SMEs) in Some Countries

No.	Country	Year	The contribution of SMEs to employment	The contribution of SMEs to GDP	The contribution of SMEs to export	SMEs contribution to total business units
1	China	2008	75%	60%	68%	99%
2	India	2012	-	-	40%	90%
3	Japan	2007	69%	53%	-	99.7%
4	South Korea	2009	75%	60%	-	99.9%
5	Singapore	-	70%	60%	-	99%
6	Malaysia	2012	56%	32%	19%	99.2%
7	Thailand	2010	-	37.1%	28.4%	99.8%
8	Philippines	2009	61.2%	35.7%	-	99.6%

The table shows that SMEs have an important role in the economic and industrial system of each country. The efforts to empower SMEs have been made through (1) improved access to funding, (2) increased productivity and competitiveness, and (3) increased exports.

2.2. Development of Accounting Information System

2.2.1. System

System is an entity or a unit which consists of interconnected parts (sub-systems), which is aimed to achieve certain goals 1996 [2]. Information is output (output) of a data processing. This output usually is in good order and has meanings for those who receive it so that it can be used as a basis for decision-making 1996 [3]. Information also means significant and useful knowledge to achieve a particular goal or objective 1993 [4]. Meanwhile, according to Henry C.Lucas, "Information is either visible or invisible fact which is available to reduce uncertainty about some circumstances or events. 2000 [5]

2.2.2. Definition of Accounting Information System

Accounting Information System is a composition of various forms, records, equipment including computers and equipment as well as communications equipment, implementation personnel, and reports which are closely coordinated and which are designed to transform the financial data to information required by the management.

The accounting system and the accounting information system have the same idea, namely: a series of administrative activities to handle corporate transactions so that they are uniform. They both are equipped with a variety of procedures, documents and journals where their results are financial reports for internal and external uses 1994 [6]. The accounting system utilizes the existing resources within the company, namely: employees, automatic machines, computers and other resources. The accounting system can be carried out manually (by utilizing the employees and support automated machines such as copy machines, calculators, and typewriters) or by computerized ones (by using computers).

System development process is a set of activities, methods, best practices, automated delivery process and equipment used by the stakeholders to develop and improve information system and software continuously 2001 [7]. The stakeholders here can be defined as ones who are interested in or attracted to the existence of an information system or to a new information system. The stakeholders can be technical or non-technical personnel. In information system, the stakeholders can be the owners, users, analysts, designers, builders, sellers of the system as well as information technology consultants.

2.2.3 Information System Development Method

Rapid Application Development (RAD) is a system development strategy that prioritizes the acceleration in the system development through extensive user involvement in the use of a construction sequence, where the circuit serves as a more effective system model (prototype).

The system development process frequently uses prototyping. This method is best used to solve misunderstanding between the users and analysts, which usually occurs due to the inability of the users to clearly define their needs

Prototyping is the rapid development and the testing of work models (prototypes) of new applications through a repeated interaction process which is frequently used by information system experts and business experts. Prototyping is also called rapid application design (RAD) because it simplifies and accelerates the design of the system 2005 [8].

Some users are difficult to express their wishes to get applications that fit their needs. These difficulties need to be resolved by the analysts to understand the users' needs and translate them into the form of a model (prototype). This model is further improved continuously to suit the users' needs.

Based on its characteristics, the prototype of a system can include low fidelity and high fidelity. Fidelity refers to the level of detail in a system.

Low fidelity prototype does not describe the system detailed enough. The characteristics of the low fidelity prototype include the following: having limited function or interaction; illustrating the concept of design and layout rather than describing the interaction model, not displaying the operation of the system in detail, demonstrating the general feel and look of the users' interface, and simply illustrating the concept of general approach.

High fidelity prototype describes the system in more detailed. This prototype has a full interaction with the users where the users can enter data and interact with the system, representing the core functions that can simulate most of the functionality of the final system, and it has a very similar appearance to the actual product.

Features that will be implemented on the prototype system can be restricted with vertical or horizontal technique. Vertical prototype contains functionality detail but only to some selected features, not on the overall features of the system. Horizontal prototype includes all the features of the users' interface, but without its staple function the horizontal prototype is only a simulation and cannot be used to do the actual work.

The process of prototyping is a process which is interactive and repetitive. It combines the traditional development cycle. The prototype is evaluated several times before the end users state that the prototype is accepted. Figure 1 illustrates the process of prototyping:

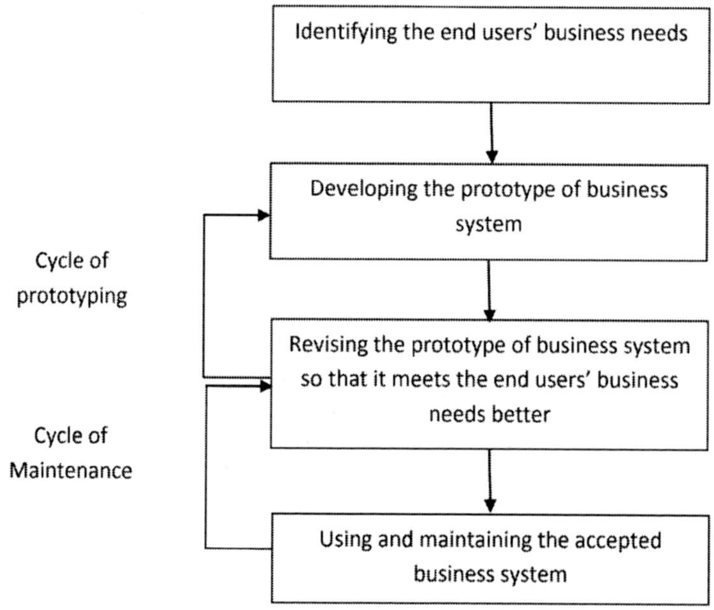

Figure 1. Stages of Prototyping

2.2.4 System Requirements Analysis

The development of information systems requires investigation and analysis of the reasons for the emergence of the idea to build and develop information system. The analysis is conducted to look at the various components used by the running system, which include hardware, software, network and human resources. The analysis also documents the activities of information system which include input, processing, output, storage and control. Furthermore, a feasibility study is conducted to formulate the required end-user information, resource requirements, costs, benefits and feasibility of the proposed project.

Analysis phase requires an evaluation system to determine the ability of the system to define what should be done by the system and then to determine the criteria to be met by the system. Several criteria that must be met include goal attainment, speed, cost, quality of information produced, efficiency and productivity, accuracy and validity and reliability 2001 [9].

2.2.5 System design

Analysis system describes what the system must do to meet the users' information needs. The system design determines how the system will meet these objectives. The system design consists of design activity that results in a functional specification. The system design can be seen as an interface design, data, and processes with an objective to produce specifications which fit product and method of the users' interface, database structure and processing and control procedures.

The system design will generate a prototype software package. A good product should include seven parts:

1) quick and easy menu features
2) input and output display
3) easily printable report
4) data dictionary that stores information on each field including the length of the field, editing in any report and used field format.
5) database with optimal key formats and record
6) appropriate display of the online query to the data stored in the database.
7) a simple structure with a programming language that allows the users to perform special processing, timing, automated procedures and others.

2.2.6 Testing the Systems

The prototype software packages are tested, implemented, evaluated and modified repeatedly until they are acceptable to the users. The system testing aims at finding errors that occur in the system and revise the system. This stage is important to ensure that the system is free of errors. The test system includes:

1) the testing of unit to test individual components independently without the other components of the system to ensure the correct operating system.
2) the testing of module which consists of interconnected-components.
3) the testing of sub-system which consists of several modules that have been integrated.
4) the testing of the system to find errors resulting from the interaction between sub-systems and interface and to validate functional and non-functional requirements.
5) the testing of acceptance with the data entry by the users and not simulation data testing.

6) documentation by recording every stage of the work from the beginning to the end of programming.

The web-based information system testing can use the techniques and methods of traditional software testing. The Web application testing includes testing the links, browser testing, usability testing, load testing, stress testing and malar testing.

The users' acceptance of the system can be evaluated by measuring the satisfaction of users to the system being tested. The satisfaction measurement system includes display, conformity with user needs, and speed and accuracy of the system to produce the desired information user. There are several models of user satisfaction measurement of the system, including the Technology Acceptance Model (TAM), End User Computing (EUC) Satisfaction, Task Technology Fit (TTF) Organizational Analysis and Human Technology (HOT) Fit Model.

One of the measurement models which has been translated into several different languages and does not show a significant difference in the measurement results is End User Computing (EUC) Satisfaction. This model emphasizes the satisfaction of users of the technology aspects which include those of content, accuracy, format, timing and ease of use of the system.

2.2.7 Implementation

Once the prototype is accepted, the implementation of the system is ready for operation and further the learning process of the new system takes place. Then it is compared with the old system in terms of technical and operational evaluation, user interactions, and information system and technology.

2.2.8 System Design Tools

System design requires equipment such as process design tools and data design tools. The former consist of a data flow diagram and a flow diagram of the system whereas the latter include entity-relationship diagram and data dictionary.

2.2.8.1. Data Flow Diagram

Data flow diagram (data flow diagram / DFD) is a chart documentation tool that uses symbols to explain a process. This chart shows the process flow of the entire system to the user and can be set according to the capability of understanding the details of the user. DFD consists of three elements: environment, processing, data flow and data storage. One advantage of using DFD is easier for users that are less mastered the field of computers to understand the system that is to be done 2005 [10].

2.2.8.2. System Flow Chart

System Flow chart is a device which is to describe the system in detail to illustrate the flow of information system and the system flow chart to illustrate the flow of the program.

3. Research Method

This research employed the action research method as to achieve its objective, namely: the development and the optimization of application of accounting information system with Rapid Application Development (RAD). In the first phase, a study was performed on the developed models. Based on the result of the preliminary research, a model of development of accounting information system with Rapid Application Development (RAD) was formulated. This research used the focused discussion group analysis technique. The FGD was held in Klaten, Central of Java, with 20 participants consist of academics, SME and Dinas UKM. Products from SME consist of food entrepreneurs, stationery, and t-shirts.

4. Result

The SME until recently have adequately thrived, especially those managed by young entrepreneurs who dare to start an independent online business. They have been challenged to create employment opportunities, not just as a job seeker force. Those above 40 years old have started to open SME since the provision of formal sector employment begins to shrink, forcing them to dare to open an independent business, in addition to hereditary business reason especially in the field of traditional food business.

The greatest constraint perceived is the limited skills of the human resources, both managers and business owners, especially in the areas of accounting and finance. In terms of accounting and finance, so far, only a manual conventional recording with a simple bookkeeping has been performed, not detailed computerized recording system assisted with detailed and accurate accounting system. Such simple recording cannot provide accurate financial information. Most SME do not understand what accounting system is and how to run it for business continuity.

For those aged over 40 years old who run SME, the procurement of computer equipment or supporting facilities is still regarded as an onerous burden of investment cost as many of them feel that such support device is not required yet, and they also find it difficult to operate the device due to their educational background. In majority, they feel more comfortable with the conventional recording system as they do today.

There has not been a separation between the operational expenses of the enterprises and the expenses for personal interests. Some owners of SME acknowledge that they still pay their personal expenses with the money of their enterprises whenever they need, and they frequently forget to record the expenses. Financial records are still complicated. It is, therefore, difficult to evaluate to what extent the SME have experienced business development because the financial records are not clear and accurate. The growth potential of the business is only judged on the basis that the enterprises "can survive and meet their need". This is mainly experienced by SME of traditional food.

The development of accounting information system for SME meets some challenges and urgencies, namely: (a) the improvement of skills of the human resources of SME in using the computerization facility, (b) the capital for procurement or investment of computer and accounting system, and (c) The mentoring of the relevant authorities so that the management of SME is sustainable, performs smoothly as expected and there is not momentum period.

5. Conclusion

The management of small and medium enterprises (SME) still uses a manual or conventional system. The skills to conduct a good accounting information system in accordance with the standard are limited. The development of accounting information system for SME encounters some challenges and urgencies. They include (a) the improvement of skills of the human resources of SME in using the computerization facility, (b) the capital for procurement or investment of computer and accounting system, and (c) The mentoring of the relevant authorities so that the management of SME is sustainable, performs smoothly as expected and there is not momentum period.

References

1. K. Mizunoura. *Accounting Standards for Small- and Medium-Sized Enterprises in ASEAN*. ERIA

   Research Project FY2015 No.22. © Economic Research Institute for ASEAN and East Asia (2017)
2. Z.Baridwan, *Akuntansi Keuangan Menengah*: BPFE (1996)
3. HM, Jogiyanto. *Sistem Informasi Akuntansi Berbasis Komputer*. Yogyakarta: BPFE (1998)
4. HM, Jogiyanto. *Pengenalan Komputer: Dasar Ilmu komputer, pemrograman, sistem* (2000) *informasi dan inteligensi buatan*. Yogyakarta: Andi Offset
5. HM, Jogiyanto. *Analisis dan Desain Sistem Informasi*. Yogyakarta: Andi Offset (2005)
6. W.W.Winarno. Sistem Informasi Akuntansi. (1994)
7. Whitten, Jeffrey L., Bentley, Lonnie D, Dittman, Kevin C. *Systems Analysis and Design Methods*, edisi ke-6. Mc.Graw-Hill, New York.(2004)
8. O'Brien, J.A. *Introduction to Information System: Essential for the E-Business Enterprise, 11th edition. Mc.Graw Hill, New York.* (2003)
9. B.Ladjamudin, Al-bahra.. *Konsep Sistem Basis Data dan Implementasinya*. Yogyakarta: Graha Ilmu (2005)
10. L.Raymond. *Sistem Informasi Manajemen*, edisi Indonesia, edisi ke-7, jilid 2. Terjemahan Teguh, Hendra dan Sukardi, Hardi. PT.Prehallindo, Jakarta. (2001)

Application of LEAP model on long-term electricity demand forecasting in Indonesia, period 2010-2025

Jaka Windarta [1,3], *Bambang* Purwanggono[2], and *Fuad* Hidayanto[1,*]

[1] Master Program of Energy, Diponegoro University, Semarang, Indonesia
[2] Department of Industrial Engineering, Diponegoro University, Semarang, Indonesia
[3] Department of Electrical Engineering, Diponegoro University, Semarang, Indonesia

Abstract. Electricity demand forecasting is an important part in energy management especially in electricity planning. Indonesia is a large country with a pattern of electricity consumption which continues to increase, therefor need to forecasting electricity demand in order to avoid unbalance demand and supply or deficit energy. LEAP (Long-range Energy Alternative Planning System) as a tool energy model and Indonesia as a case study. Basically, electricity demand is influenced by population, economy and electricity intensity. The purpose of this study is to provide understanding and application of electricity demand forecasting by using LEAP. The base year is 2010 and end year projection is 2025. The scenarios of simulated model consist of two scenarios. They are Business as Usual (BAU) and Government policy scenario. Results of both scenarios indicate that end year electricity demand forecasting in Indonesia increased more than two fold compared to base year.
Keywords: electricity, demand forecasting, LEAP, scenario, Indonesia.

1 Introduction

Energy has an important role in human life since ancient times. Animals were used to plow rice fields, firewood for cooking and the use of wind for sailing is proof that energy is important in supporting human life [1]. Energy sources, in terms of usage, consist of primary energy and secondary energy. Primary energy is the energy given by nature directly without processing. While secondary energy is the primary energy that has further process. Petroleum, natural gas and coal are forms of primary energy. Electricity is one form of secondary energy. Based on the type of fuel, electricity is the most widely used energy after petroleum [2]. In addition, electricity is easily to trasmitted through transmission and distribution line, easy to use and a lot of devices on the world need electricity as power source.

Indonesia is the fourth most populous nation in the world. The archipelago approximately 17,508 islands located between latitude 6° North and 11° South and longitude 95° East and 141° East [3]. As reported in the 2010 census conducted by Statistics

* Corresponding email: fuad.hidayanto@yahoo.com

Indonesia (Badan Pusat Statistik, BPS), the population of Indonesia was 237.5 million with population growth of 1.37%. GDP growth per annum from 2010 to 2015 average increase of 5.72% and reach highest value of 6.5% [4]. GDP growth in 2015 decreased only 4.57%, it is due to global economy not stable affecting demand of Indonesia's export product. There is a relationship between economic growth and electricity consumption in Indonesia [5].

The share of electricity consumption is 18.2% of the national energy consumption in 2015 [4]. Electricity demand in the future will increase along the increasing number of electrical equipment products. For example electric stoves, electric bikes and electric cars have began to be developed. Electrical equipment is considered more environmentally friendly than existing equipment with fuel oil and gas that may cause air pollution. The increasing growth of electricity demand need to be anticipated with the provision of power plants. In the electricity management need planning of electricity demand and supply of power plants.

Base on the reveal above, electricity demand forecasting in future is important in order to avoid unbalance demand and supply or deficit energy. Several methods have been developed such as the markov model, fuzi linear regression and a combination of existing methods [2]. Factors or variables that affect electricity demand include population, gross domestic product (GDP), electrification ratio and other intensity.

2 Literature Review

Electricity supply in Indonesia shall be under the control of the state, which supplies shall be provided by the government and the regional government under the principle or regional autonomy. The implementation of electricity supply business is carried out by state-owned enterprises (BUMN), regional-owned enterprises (BUMD), private enterprises, cooperatives and self-supporting communities [6]. Every power supply entrepreneur is obliged to hold a business license for electricity supply (IUPTL). PLN is one of the holders of IUPTL which business area covers the entire territory of the Republic of Indonesia as figure 1.

In 2015, the installed capacity of power plants in Indonesia reaches 55.5 thousand MW consisting of PLN power plants amounting to 38.3 thousand MW and non PLN of 17.2 thousand MW [7]. Compared to 2014 the installed capacity of power plants has been risen by 4.64%. The supply of electricity amounted to 233,981.98 GWh which consists of PLN's electricity production of 176,472.21 GWh and purchases of 57,509.77 GWh. Electricity consumption in 2015 amounted to 202,845.82 GWh, an increase of 2.13% from the previous year. Table 1 present the electricity consumption in Indonesia.

PLN customers consist four main sectors. They are household customers, industrial customers, business customers and public customers. The number of customers in year 2015 reached 61,167,980 customers increasing of 6.39% over the previous year [8]. From the total number of customers, the household group is the largest number of customers, 56,605,260 customers or 92.54%. The electrification ratio is the ratio of the electrified household to the number of households. Electrification ratio in 2015 reached 88.30% or increased 3.94% from the previous year [7].

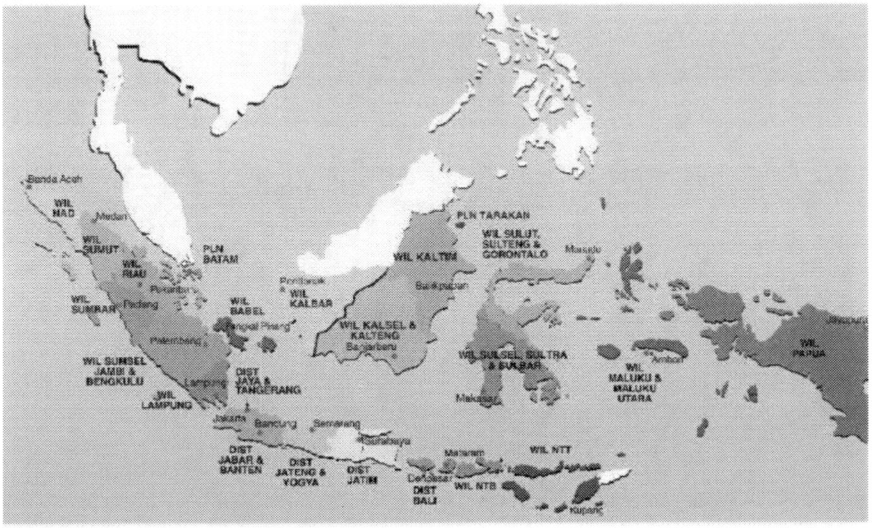

Fig. 1. Map of Business Area PLN [9]

Table 1. Electricity Consumption in Indonesia [8]

Sector	Electricity Consumption (GWh)					
	2010	**2011**	**2012**	**2013**	**2014**	**2015**
Household	59.824,94	65.111,57	72.132,54	77.210,71	84.086,46	88.682,13
Industry	50.985,20	54.725,82	60.175,96	64.381,40	65.908,68	64.079,39
Business	27.157,22	28.307,21	30.988,64	34.498,38	36.282,42	36.978,05
Public	9.330,11	9.848,06	10.693,60	11.450,53	12.324,21	13.116,25
Total	147.297,47	157.992,66	173.990,74	187.541,02	198.601,77	202.855,82

3 The Mothodology of Research

3.1 LEAP Model

The LEAP (Long-range Energy Alternative Planning System) model is a static energy - economicy - environment model developed by the Stockholm Environment Institute in Boston, USA since 1980 [10]. LEAP has been used by more than 190 countries in the world [2]. The Leap model can be used to calculate energy requirements, transformations and energy sources.

The modeling methodology in LEAP is accounting. Energy demand or energy supply is calculated by summing up the energy consumption and supply of each type of activity. Leap provides four main modules used in energy modeling. They are key assumptions, demand, transformation, and resources [11]. Key assumptions module consists of data such as population size, population growth, nominal GDP (Gross Domestic Product), GDP growth, number of household customers and other parameters. Demand module consists of four sectors, they are household, industry, business and public.

3.2 Final electricity demand analysis

Final electricity demand analysis is calculated for each year and branch (sector). Electricity demand is calculated as a product of total activity and electrical intensity [2].

$$ED_{b,s,t} = TA_{b,s,t} \cdot EI_{b,s,t} \tag{1}$$

Where ED is electricity demand, TA is total activity, EI is electricity intensity, b is sector, s is scenario and t is time (year). In the final electricity demand analysis, energy intensity can be calculated from electricity consumption and the number of customer [2]. It can also be energy intensity defined as the total amount of energy consumption per unit of gross domestic product (GDP) [12]. The number of customer and GDP is referred to as total activity.

$$EI_{b,t} = \frac{EC_{b,t}}{TC_{b,t}} \tag{2}$$

or

$$EI_{b,t} = \frac{EC_{b,t}}{GDP_{b,t}} \tag{3}$$

Where EI is electricity intensity, EC is electricity consumption, TC is total customer number, GDP is gross domestic product, b is sector and t is time (year).

3.3 Forecasting method

The electricity demand forecasting in this study is calculated by a very simple model. For household sector, the electricity demand is calculated from the number of household customers and its electricity intensity. Projected number of household customers based on population growth, household size (family member) and electrification ratio.

$$TC_{h,s,t} = \frac{P_{s,t}}{H_{s,t}} \cdot ER_{s,t} \tag{4}$$

Where TC_h is total customer of household sector, P is number of population, H_S is average of household member (household size) and ER is the electrification ratio. The number of households is calculated from the division of number population with household size. Electrification ratio is ratio of number of households with electricity divided by number of households throughout Indonesia or the percentage of households with electricity [12].

$$ED_{h,s,t} = \frac{P_{s,t}}{H_{s,t}} \cdot ER_{s,t} \cdot EI_{h,s,t} \tag{5}$$

Where ED is electricity demand in household sector, EI is the electricity intensity in household sector, P, HS and ER, s, t are as defined above. So the electricity demand forecasting for household sector can be calculated by projecting population, household size, electrification ratio and electricity intensity [2].

Electricity demand for non-household sector is calculated simpler than household sector. There are only two factors used in this method. They are gross domestic product and electricity intensity.

$$ED_{i,s,t} = PDB_{i,s,t} \cdot EI_{i,s,t} \tag{6}$$

$$ED_{b,s,t} = PDB_{b,s,t} \cdot EI_{b,s,t} \quad (7)$$

$$ED_{p,s,t} = PDB_{p,s,t} \cdot EI_{p,s,t} \quad (8)$$

Where ED is electricity demand while i, b dan p describe branch or sector, business and public.

3.4 Scenario Developments

In this study, the base year of the leap model is set from 2010-2015. The fist year scenario start in 2016 and ends in 2025. The population, household size (family member) and GDP at constant price 2000 from 2010-2015 as Table 2. The electrification ratio is set at 100% by 2020 according to Government targets [12][13].

Household electrical intensity used average value from 2010 to 2015 which is 1,549 kWh per household customer. In the non-household sector, energy intensity for each industry, business and public sectors was 22.36, 12.02 and 4.16 kWh per million rupiah, respectively.

Table 2. Population, Household size and GDP [4]

Indication	Year					
	2010	**2011**	**2012**	**2013**	**2014**	**2015**
Population (Thousand People)	237.641	238.519	245.425	248.818	252.165	255.462
Household size	3,89	3,84	3,89	3,89	3,89	3,90
GDP at Constant Price 2000 (Trillion Rupiahs)	2.314	2.465	2.618	2.770	2.909	3.042

Two scenarios will be simulated in this study. They are Business as usual (BAU) scenario dan Government policy scenario. BAU scenario used trend data reference from year 2010-2015. Government policy scenario is influenced government policy such as BPS and RPJMN data.

3.4.1 Business as Usual (BAU) scenario

In BAU scenario the population growth projection uses the average value of population growth from 2010-2015 is 1.44%. Household size using the average value of household size from 2010-2015 is 3.9. Projected economic growth using the average GDP growth from 2010-2015 amounted to 5.72%.

3.4.2 Government policy scenario

In government policy scenario the population growth projection uses the Indonesia population projection data 2010-2035 [14]. For the period of 2010-2015 the population growth projection is 1,19% and for the period of 2020-2025 is 1,00%. Projection of economic growth in 2015-2019 adopted economic growth rate in RPJMN year 2015-2019 issued by Bappenas that is equal to 7,1%. For the period of 2020-2025, the economic projection uses the projected number of BPPT, which is an average of 6.4% per year [9].

4 Results and Discussion

Electricity demand forecasting for the BAU scenario and the Government's policy scenario are shown as in table 3 and table 4, respectively. In both scenarios, at the end of the scenario there is an increase of more than double the electricity requirement of the base year. In the 2025 BAU scenario, electricity needs to be 322.0 GWh or an increase of 174.7 GWh from 2010. In the Government's policy scenario, electricity demand in 2025 is higher at 337.1 GWh or an increase of 189.9 GWh from 2010.

In the household sector the BAU scenario of electricity demand projection is higher than the Government scenario. This is because the population growth BAU scenario is higher, probably due to Indonesian culture that states many children a lot of fortune. In the industrial sector, business and public BAU scenarios of projected electricity demand are lower than the Government scenario. This is because the economic growth as input reference scenario of BAU is the year 2010-2015 is still low due to global economy not stable.

Overall projection of BAU scenario electricity requirement is lower than Government policy scenario as shown in Figure 2. It is because of the optimism of the Government in increasing economic growth.

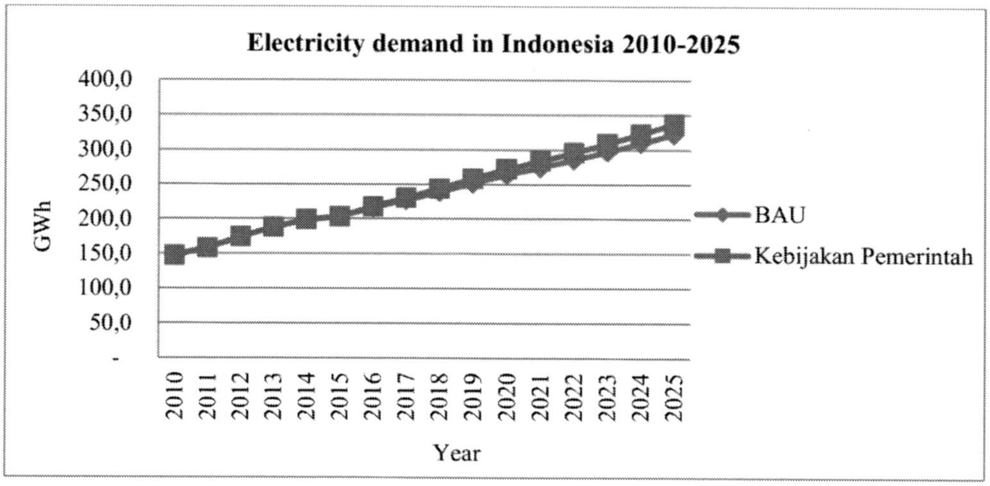

Fig. 2. Electricity demand in Indonesia 2010-2025

Table 3. Electricity Demand in Indonesia 2010-2025, Scenario BAU

Sector	Electricity demand (GWh)			
	2010	2015	2020	2025
Household	59,8	88,7	109,5	117,7
Industry	51,0	64,1	89,8	118,6
Business	27,2	37,0	48,3	63,8
Public	9,3	13,1	16,6	21,9
Total	147,3	202,9	264,3	322,0

Table 4. Electricity Demand in Indonesia 2010-2025, Scenario Government Policy

Sector	Electricity demand (GWh)			
	2010	2015	2020	2025
Household	59,8	88,7	108,0	113,5
Industry	51,0	64,1	95,2	129,8
Business	27,2	37,0	51,2	69,8
Public	9,3	13,1	17,6	24,0
Total	147,3	202,9	272,0	337,1

5 Conclusion

By the increasing of the population and economy every year, influence of increasing the amount of electricity consumption in Indonesia. Without anticipating the provision of proper power generation capacity can cause serious problems such as shortage of power supply. Therefore, the projection of electrical demand becomes important in planning the addition of generating capacity in the future. This study analyzes electricity demand forecasting from 2010-2025. Factors used in this projection are population, household size, electrification ratio, number of household customers, GDP and electricity intensity. The electricity demand forecasting is calculated by a simple mathematical model based on the value of the input variable. Simulated scenarios are two BAU scenarios and Government policy scenarios. BAU scenarios of data inputs use data averages of 2010-2015 while government policy scenarios use government targeted values such as BPS and RPJMN data. The results show an increase in electricity demand more than two fold by 2025 when compared to 2010. The projection of electricity demand for government policy scenarios is higher than the BAU scenario. This is because Government policy scenario optimism in improving economic growth, while BAU scenario in the year 2010-2015 as the reference year of a slowdown in the national economy due to the global economic crisis.

6 Recommendation

This study calculates only the projection of electricity demand, while for additional capacity of the electricity power plants has not been discussed. Therefore, further research on the readiness of electricity power plants is needed to evaluate their capacity to meet the growing demand for electricity in the future.

References

1. P. Yosgiantoro, *Energy Economy. Theory and Practice*, Pusaka LP3ES Indonesia, Jakarta (2000)
2. Suhono and Sarjiya, *Long-Term Electricity Demand Forecasting Of Sumatera System Based On Electricity Consumption Intensity And Indonesia Population Projection 2010-2035,* Energy Procedia **68**, 455-462 (2015)
3. H. Batih and C. Sorapipatana, *Characteristics of urban households' electrical energy consumption in Indonesia and its saving potentials*, Renewable and Sustainable Energy Reviews **57**, 1160-1173 (2016).
4. Pusdatin ESDM, *Handbook of Energy & Economic Statistics of Indonesia,* (2016)

5. T.D. Hartono and M. Kuncoro, *Electricity consumption and economic growth: A Causality Evidence from Six Economic Corridors of Indonesia*, Kuala Lumpur International Business, Economics, and Low Conference KLIBEL, 1984-2010 (2013)
6. Government of Republic of Indonesia, *Law of the Republic of Indonesia number 30 Year 2009 Concerning Electricity*, (2009)
7. Directorate General of Electricity, Ministry of Energy and Mineral Resourses, *Electricity Statistics 2016*, Jakatra, (2016)
8. PT PLN, *PLNs Statistics 2015*, (2015)
9. PT PLN, *RUPTL PT PLN (Persero) Tahun 2016-2025*, (2016)
10. T.O. Winarno, *LEAP Energy Planning Guide*. Center for Energy Policy Studies. Institut teknologi Bandung, (2006)
11. U. Perwez, A. Sohail, S.F. Hassan and U. Zia, *The Long-Term Forecast of Pakistan's Electricity Supply And Demand: An Application of Long Range Energy Alternatives Planning*. Energy **93**, 2423-2435 (2015)
12. Government of Republic of Indonesia, *Government Regulation of Republic of Indonesia No. 79 Year 2014 regrading National Energy Policy*, (2014)
13. DEN, *Indonesia Energy Outlook*, Jakarta (2016)
14. Statistics Indonesia (BPS), *Indonesia Population Projection 2010-2035*, (2013)

Review on active suspension system

Aizuddin Fahmi Mohd Riduan[1], *Noreffendy* Tamaldin[2], *Ajat* Sudrajat[3], and *Fauzi* Ahmad[4]

[1,2,4]Centre for Advanced Research on Energy, Universiti Teknikal Malaysia Melaka, Hang Tuah Jaya, 76100 Durian Tunggal, Melaka Malaysia.
[1,2,4]Faculty of Mechanical Engineering, Universiti Teknikal Malaysia Melaka, Hang Tuah Jaya, 76100 Durian Tunggal, Melaka, Malaysia.
[3]Engineering Physics, Faculty of Science and Technology, Universitas Nasional-Jakarta JL. Sawo Manila, Pasar Minggu, Jakarta 12520, Indonesia.

Abstract. For the past decade, active suspension systems had made up most of research area concerning vehicle dynamics. For this review, recent studies on automobile active suspensions systems were examined. Several vehicular suspension types were also described to compare amongst them. From published investigations by previous researchers, various automotive suspensions in terms of cost, weight, structure, reliability, ride comfortability, dynamic and handling performance were exhibited and compared. After careful examination, it was concluded that electromagnetic active suspensions should be the general direction of vehicle suspension designs due to its energy regeneration, high bandwidth, simpler structure, flexible and accurate force control, better handling performance as well as drive characteristics.

Keywords: Active Suspension, Handling Performance, Dynamic Performance

1 Introduction

Vehicle suspension main task is to separate passenger and vehicular body interactions from oscillations generated by road abnormalities whilst still maintaining continuous wheel-road contact. Currently, there are three types of automotive suspensions commonly used namely passive, semi-active, and active.

All said systems are based by either pneumatic or hydraulic operation. It was asserted that some of these suspension systems cannot fully solve automobile oscillations problem because they are very costly and lend towards vehicle energy consumption increment [1].

Active vehicle suspension systems had been in wide investigation for more than 20 years due to their promising characteristics. These systems poses potentiality of responding considerably good towards upward and downward changes caused by road inputs irregularities since the dampers as well as springs are mediated by an actuator force. Actuator in an active suspension functions to spread energy away from system and it can be operated through various kinds of controllers determined by proposed design. With correct managing techniques which could induce more beneficial compromise between driving comfort and road handling stability, an active system as a whole would be the superior suspension design [2].

Sophisticated electromechanical and electronic systems are used to affect automobile dynamic performances such as anti-lock braking system (ABS), electronic stability control (ESC), electronic brake force distribution, and many more. These systems advancement towards supporting passenger safety and vehicle handling have become an increasing requirement in the automotive industry, especially when vehicles nowadays can vary in design scopes from either being small, having higher centre of gravity, or aim to be environmental friendly.

One other present trend in the automotive industry is movement towards fully electric vehicles. These cars possess different overall weight distribution due to an internal combustion engine (ICE) is substituted with an electric motor and a battery pack that weighs more or less 400 kg. Efficiency of electrical drive train is best optimized when it is fully incorporated to the wheel. Even so, car unsprung ass will gain which could cause problems with respect to passenger comfort and handling. Hence, this clearly shows the need for active suspension be integrated in automobiles [3].

Automotive manufacturers often compete against each other to develop dependable and comfortable drive characteristics while at the same time also producing best stability during braking and cornering. With passive suspension however, it could just provide one of the stated features or at most compromise between handling and ride comfortability. Nevertheless, selected companies do offer adjustable suspension system options which can be tuned according driver comfortability needs. Because of this, researchers had proposed and employed several semi-active and active vehicle suspension systems both in theoretical and experimental terms [4]. These systems main benefit is that they simultaneously permits eliminating most road interferences from pitch control as well as active roll.

2 Classifications of automotive suspensions

2.1 Passive Suspension

Passive vehicle suspension systems modelled in **Fig. 1** is a typical mechanical suspension arrangements comprising viscous damper and linear springs with constant damping coefficient and stiffness respectively [2].

Passive vehicle suspension is reliable, not complex, as well as inexpensive. The damper and spring in this system are fastened between car body and wheel support structure. A damper insides is occupied with a hydraulic oil or compressed gas, and there is a piston moving by a rod from its exterior. Piston movement is permitted by a hole that allows fluid passing among cylinder parts. This flow of fluid develops a reactional force which is relative towards proportional speed between unsprung and sprung masses. Damping is then attained by converting oscillations energy into heat.

This uncomplicated system cannot yield decent outcomes for counteracting suspension problems since it does not give any external control and feasible betterments in terms of materials, hole valves, or even shapes [1].

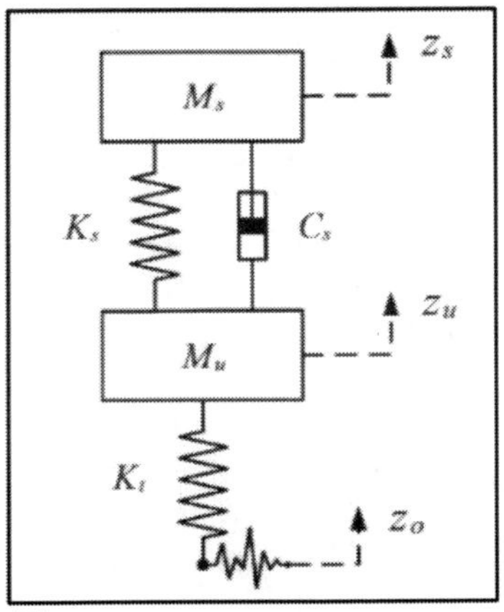

Fig. 1. Passive suspension model [6]

2.2 Semi-active Suspensions

Semi-active suspension is similar to a passive design with only difference being having a variable damping coefficient but still fixed spring constant and also absence in active force sources. This kind of system allows seamless change between a passive damper with semi-active damping coefficient as shown in **Fig. 2**.

While maintaining fixed spring utilization, the employed continuous variable damper alters workability in real time through closed feedback loop control design to assure proper energy dissipation, thus bettering suspension overall performance against a passive system [5].

A semi-active suspension can be remotely switched electrically to stiffen or soften the system making its damping coefficient either varies discontinuously or continuously. Shifting in functioning strategies like stiffening suspension would benefit vehicle at cornering, braking, acceleration, reducing low frequency response towards inertial forces associated with roll and pitch, as well as avoiding wheel resonances and body excitation.

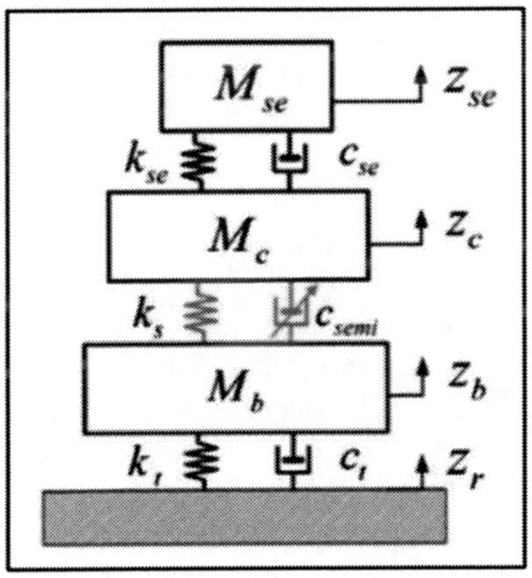

Fig. 2. Semi-active model [7]

2.3 Active Suspensions

Compared to other previous systems, an active suspension includes an actuator that can

 https://doi.org/10.1051/shsconf/20184902008

supply active force regulated by a controlling algorithm which uses information gathered from attached vehicle sensors. As illustrated in **Fig. 3**, active suspension comprises an actuator, mechanical spring, and damper; or an actuator and mechanical spring only.

These kind of systems have much better reacting capabilities against generated vertical forces caused by unpredictable road input changes since the dampers as well as springs are regulated through an actuator force. This actuator operates by allowing or spreading energy from the system and could moderate through different controller types based on intended design. With proper controlling methods, an active suspension can result in compromise between vehicle ride comforts to road handling stableness be more improved, thus making it an overall enhanced suspension design.

Because of this, most researchers recognized that this system type are among best ways in improving its overall performance due to the ability in managing contradictory parameters [5]. Even so, it is a complicated and expensive system which explains this particular system usage by only a small number of high-end car models, or trucks [1].

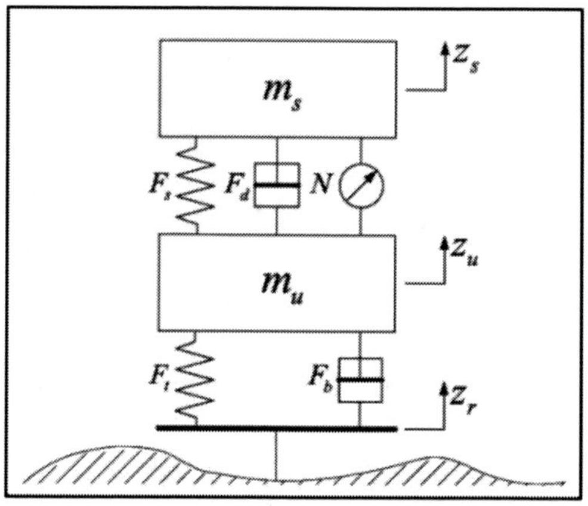

Fig. 3. Active model [8]

3 Hydraulic or pneumatic active suspensions

Active suspension system with pneumatic or hydraulic actuator are managed through electric drives and the power is provided from a battery source or conventional ICE itself. Typically, hydraulic systems are more widely used in body control systems due to its design simplicity, high force density, technology maturity, reliability, as well as various commercial parts availability.

For example, BMW recently developed an anti-roll control system by putting a hydraulic rotary actuator at rear of vehicle anti-roll bar centre. Another instance was exemplified by Mercedes active body control system which applies high pressure hydraulics to pre-stressing the springs, therefore developing anti-roll forces without coupling right and left wheels [2]. However, there are disadvantages in using a hydraulic system such as:

a. Inefficiently performed because continuously requiring a pressurized system
b. Comparatively high system time constant (flexible hoses and pressure loss)
c. Environmentally damaging due to toxic hydraulic fluids presence caused by hose leaks and ruptures
d. System volume and infeasibility space requirements even though it primarily contribute towards overall sprung mass

4 Electromagnetic active suspensions

An electromagnetic active suspension system consists a spring and electromagnetic actuator setup in parallel arrangements within unsprung and sprung mass. The electromagnetic actuator performs with regards to electric supplied range by implemented control systems. It brings about active controlled forces for speedily absorbing road shocks, suppressing pitch and roll motions, and improving comfort as well as safety. Bi-directional amplifier electromagnetic actuator on the other hand, works by yielding and reducing electric power consumption compared against a hydraulic actuator [2]. After comparing with hydraulic actuators, electromagnetic actuators main superiority are as bellow:

a. Efficiency increment
b. Dynamic behaviour improvement
c. Stability betterment
d. Force control precision
e. Dual actuator operation

The weak points meanwhile are:

a. Suspension volume increment
b. Relatively greater current requirement from a 12 to 14 V system
c. Conventional designs which requires excitation to supply continual force
d. More expensive of a system

5 Comparisons

Table 1 below summarized suspension system comparisons with respect to descriptive classification of reviewed vehicular suspensions.

Table 1. Comparison between various automotive suspension systems

Parameters	Passive suspensions	Semi-active suspensions	Hydraulic/pneumatic	Electromagnetic active
Structure	Simplest	Complex	Most Complex	Simple
Weight/Volume	Lowest	Low	High	Highest
Cost	Lowest	Low	Highest	High
Ride comfort	Bad	Medium	Good	Best
Handling	Bad	Medium	Good	Best
Reliability	Highest	High	Medium	High
Dynamic	Passive	Passive	Medium	Good

6 Recent innovations on suspension

6.1 Magnetic dampers

By altering electric current through a ferrous fluid, Magnetic Ride Control (MRC) dampers can adapt its stiffness in response towards specific driving situations. Even though this system is licensed to other car manufacturers, General Motors (GM) which first designed it is still leading in competition. Its third generation MRC as shown in selected performance vehicles like the Chevrolet Corvette Stingray **(Fig. 4)**, amended former designs with a

 https://doi.org/10.1051/shsconf/20184902008

second wire addition. With that component add-on, suspension fluid could interchanged from firm to standard mode actively, whereas previously lag would be present when particles regained back to their normal state.

Fig. 4. Electromagnetic suspension [9]

6.2 Active Curve Tilting

This technology was developed in 2015 by Mercedes-Benz and its implementation in the S65 AMG Coupe utilizes a lateral acceleration sensor paired with a forward looking camera to perceive corners whilst also employing air suspension to tilt its body in towards the apex. As displayed in **Fig. 5**, this application was not aim for increasing cornering performance but instead enhancing ride comfortableness as passenger would undergo lower lateral loads at a given speed. Essentially, if road surface does not have a well banked corner, the vehicle would simulate one for its occupants.

Fig. 5. Active Curve Tilting suspension [10]

 https://doi.org/10.1051/shsconf/20184902008

6.3 Hydraulic Roll Control

Audi RS 5 shown in **Fig. 6** is one of the recent automobiles to feature a hydraulic cross-linked suspension which Audi designated as Dynamic Ride Control. Unlike mainstream suspensions which applied steel anti-roll bars for counteracting body roll, this hydraulic system would transfer fluid towards the vehicle opposite side. It is as if possessing a huge anti-roll bar but without any at all, depending on driving conditions. Cars with best balance between handling and performance would to some extent use variations of this approach. Even though hydraulic cross-linking is not entirely perfect, but it is currently the nearest suspension technological design with regards to a fully active suspension.

Fig. 6. Hydraulic Roll Control suspension [11]

7 Conclusions

Published work concerning on vehicle suspensions studies and its recent developments had been looked upon thoroughly in this paper.

Based on examined findings, electromagnetic active suspensions was postulated to be the future tendency of automotive suspension designs due to its energy regeneration, not complex structure, high bandwidth function, flexible and accurate force control, good ride quality as well as handling performance.

In recent years, advancement in power electronics, microelectronics, and permanent magnet materials had permitted significant betterments in electrical drives domain.

Steady and dynamic state performance, weight and volume decrement, electronic control system unconstrained incorporation in cars, reliability, as well as downsizing are crucial factors for rationalizing towards more generalized electrical drive utilizations.

The progression in this technology field warrants analysis of possibility enforcing available suspension systems with electromagnetic actuators to increase its performance without raising costs and energy consumption.

Succeeding investigations in electromagnetic active suspensions should concentrate on two main prospects which are research and developing more towards electromagnetic linear actuators as well as better management of the system.

Acknowledgement

The authors would like to acknowledge Center of Advance Research on Energy (CARe), Universiti Teknikal Malaysia Melaka (UTeM) and Engineering Physics Department, Faculty of Science and Technology, Universitas Nasional-Jakarta for their continuous support in this study.

References

1. I. Martins, J. Esteves, F. P. da Silva, P. Verdelho, *Electromagnetics hybrid active-passive vehicle suspension system*, Technical University of Lisbon, Lisbon, Portugal (2015)
2. M. Mohammed Bello, A. Y. Babawuro, S. Fatai, *Active suspension force control with electro-hydrolic actuator dynamics,* ARPN Journal of Engineering and Applied Sciences, **10**, 23, pp.17327 – 17331 (2015)
3. B. L. J. Gysen, J. L.G. Janssen, *Active Electromagnetic Suspension System for Improved Vehicle Dynamics*, IEEE Transactions On Vehicular Technology, **59**, 3, pp.1156 – 1163 (2016)
4. R. Rosli, M. Mailah, G. Priyandoko, *Active Suspension System for Passenger Vehicle using Active Force Control with Iterative Learning Algorithm*, WSEAS Transactions on Systems and Control, **9**, 2, pp.120 – 127 (2014)
5. M. Shafie, M. Bellob, R. M. Khan, *Active Vehicle Suspension Control using Electro Hydraulic Actuator on Rough Road Terrain*, Journal of Advanced Research in Applied Mechanics, **9**, 1, pp.15 – 30 (2015)
6. M.M. Moheyeldein, A.M.A. El-Tawwab, K.A.A. El-gwwad, M.M.M. Salem, *An analytical study of the performance indices of air spring suspensions over the passive suspension*, Beni-Suef University Journal of Basic and Applied Sciences (to be published)
7. K. Sim, H. Lee, J.W. Yoon, C. Choi, S.H. Hwang, *Effectiveness evaluation of hydro-pneumatic and semi-active cab suspension for the improvement of ride comfort of agricultural tractors*, Journal of Terramechanics, **69**, pp. 23 – 32 (2017)
8. H. Pan, W. Sun, X. Jing, H. Gao, J. Yao, *Adaptive tracking control for active suspension systems with non-ideal actuators*, Journal of Sound and Vibration, **399**, pp. 2 – 20 (2017)
9. Motor Trend: 2014 Chevrolet Corvette Stingray Z51 First Test. [online] Available at: https://www.motortrend.ca/en/news/2014-chevrolet-corvette-stingray-z51-first-test/#2014-chevrolet-corvette-stingray-z51-suspension [Retrieved 1st August 2018]
10. Popular Mechanics: 3 Technologies That Are Making Car Suspensions Smarter Than Ever. [online] Available at: https://www.popularmechanics.com/cars/car-technology/a14665/why-car-suspensions-are-better-than-ever/ [Retrieved 1st August 2018]
11. Audi Technology Portal: Dynamic Ride Control. [online] Available at: https://www.audi-technology-portal.de/en/chassis/suspension-control-systems/dynamic-ride-control_en [Retrieved 1st August 2018]

Simulation and economic analysis of solar cooling for building in tropical climate of Surabaya, Indonesia

Elieser Tarigan[1,*]

[1]Electrical Engineering Department and PSET, Faculty of Engineering, University of Surabaya Jalan Raya Kalirungkut, Surabaya, Indonesia

Abstract. The possibility of solar cooling technologies is simulated and discussed in this work. Cooling system application for a six-floor university building in Surabaya Indonesia was taken as a case study. Two different solar technologies systems were designed and compared: (i) photovoltaic powered cooling system, and (ii) solar thermal absorption cooling system. Economic analysis was carried out based on the economic key-figures as well as the CO_2 emission analysis. Based on the results gained in the profitability analysis, the most economically feasible system is solar thermal absorption cooling system due to the good agreement between solar radiation and cooling demand. Besides, this systems allows the maximum CO_2 emissions savings.

1 Introduction

Air conditioning for hot climate conditions is a growing market due to increased thermal loads, as well as improving living standards. There are now different cooling technologies available on the market used to provide temperature and humidity control on premises. The most popular conditioning systems use the compressor cycle in order to gather the surplus heat from indoors and release it outdoors. The refrigerant is a working fluid used to absorb, transport and release heat. The whole process is driven by electricity. Absorption chillers are regaining popularity because of the possibility that they offer to use residual thermal energy in combined heat and power applications. Moreover, they can be successfully applied in solar cooling systems [1].

Solar cooling technology, especially in hot climates, has huge energy-saving potential in comparison to conventional technologies. A number of studies have been carried out on this topic. A solar absorption cooling system was simulated in TRNSYS for Spanish weather conditions by Cassals, 2006 [2]. Ali et al., (2008) [3] reported that in 2002, an integrated free cooling and solar-powered single-effect lithium bromide-water absorption chiller was built in Oberhausen, Germany. The free cooling percentage was evaluated to be up to 70%, and it covered 25% of the total cooling demand during its five years in operation. The monthly average solar fraction was calculated at between 31.1% and 100%. According to

* Corresponding author: elieser@staff.ubaya.ac.id

the calculations performed in work under discussion, the specific collector must be 4.23 m^2/kW cold to harvest the maximum possible solar heat fraction.

The potential for extending the usability of solar systems not only to cooling buildings but also to heating them in the cold months was evaluated [4]. The study showed that the most feasible uses of integrated solar cooling and heating systems are for residential houses and office buildings. The reasonable payback time can be achieved only for Southern European locations (e.g., Rome). Therefore, in order to make such technology more competitive, the prices for solar collectors and absorption chillers must be reduced.

In present work, two different possibilities of solar cooling technologies are simulated using Polysun software each for: (i) photovoltaic powered cooling system, and (ii) solar thermal absorption cooling system. Taking into account the high level of daily average solar radiation and the number of peak sun hours during the year the two technical solutions were considered have a good potential. The objective of the study is to find the most effective solar based technological solution for a cooling system for a building under tropical, climate of Surabaya, Indonesia. Economic analysis was carried out based on the economic key-figures as well as the CO_2 emission analysis.

2 Methods

2.1 Simulated building characteristics

The object of the present study is an existing six-floor library building at the University of Surabaya, Indonesia. The building measures 10,625 m2 in total. The front of the building faces South West. The front and top views, taken from Google Earth™ [5] of the building are shown in Figure 1.

Fig. 1. Front and top view of the simulated building

The building consists of two parts namely building A and building B. Inside of the building some stairs connecting the floors and building A and B. Outside of the building consist of:

•terrace wall which is widely open

•terrace with 2.5 m width, around the building for each floor.

•main wall of the building with window-to-wall ratio (WWR) of 30%

The wall is constructed by common brick and plaster (inside), and plaster + ceramic (outside), so that the total thickness of the wall is about 160 mm. The floor is from of 20 cm x 20 cm ceramic. About 30% of façade and all sides of the exterior wall are closed windows glass of single layer with aluminum frame. Inside of the building is divided into several partitions for different purposes, such as bookshelves, reading rooms, computer rooms, offices, bathrooms, etc.

2.2 Cooling Load

When the building was constructed, the cooling system was designed with the capacity that it would supply cooling for the whole building. The system consists of 6 chillers /and compressors, each 57.42 kW, so in total 57.42 kW x 6 = 344.52 kW (electricity power). The number of power is approximately the same when it is calculated based on so-called "rule of thumb" AC sizing in Indonesia [6]

$$P = (L \times W \times H \times Factor1 \times 37) + (Poeple \times Factor2)\ [BTU]$$

where: W = width of the room (m), L = length of the room (m), H = height of the room (m), and People = number of people in the room. Factor 1 corresponds to: bad room= 5, office and living room = 6, and restaurant, salon or mini market = 7. Factor 2 corresponds to: adult = 600, and Child = 300. For the library in this study, the number of adults is assumed about 300 people.

2.3 Simulation

To simulate the system and for optimal design configuration, the simulation software Polysun is used. The software allows for dynamic simulation of the energy systems. The simulation was designed based on the given boundary conditions (Indonesian climate data and cooling demand) [7]. The cooling load was calculated based on the rule of thumb formulae applicable to this particular building type in Indonesia, i.e 344 kW.The cooling systems were compared based on energy and economic criteria. For the economic calculations, the embedded Polysun model is used. Based on the comparative analysis, the optimal solution is recommended. Two different solar technologies systems were simulated, each for: (i) photovoltaic powered cooling system, and (ii) solar thermal absorption cooling system.

3 Results and Discussion

3.1 PV powered air conditioning

The main idea of this system is to improve and enlarge the existing conditioning system so that it covers demand since according to data provided, the existing system is currently insufficient to cover the demand. Moreover, the photovoltaic system can be built using the resources saved from the existing facilities. This option is also economically reasonable because the cooling demand coincides with the time that solar energy is generated. The main components systems are PV panel and inverter with the configuration is as shown in Figure 2.

 https://doi.org/10.1051/shsconf/20184902009

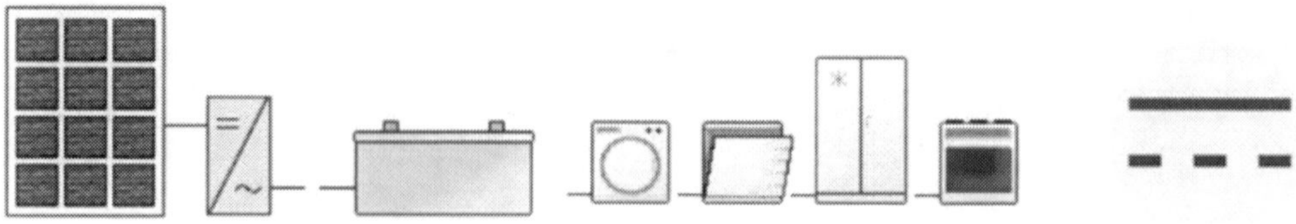

Fig. 2. The air conditioning system powered by PV system(generated from Polysun simulation software)

Simulation results show that about 376 kWp of PV panels required to full fill energy for cooling load. For the installation of the solar panel, the following layouts were considered: The library building roof was used for the photovoltaic field. However, due to the big load, the roof area is not enough to cover the electricity demand, and thus the second variant for the layout of the PV modules was considered. The PV modules are placed on the roof of the parking lot in front of the library. Of course, the main condition for this layout is the building of the carport. Another advantage of the carport is preventing cars from overheating in the sun.

3.2 Absorption Chiller With Solar Collectors

The screenshot of the system is shown in Figure 3. In the system, the collector system is used as the main energy generation. While the gas boiler covers peak loads and provides a backup auxiliary energy source (it covers periods during which solar radiation is not sufficient, such as cloudy days or mornings when the charge in the water storage is not

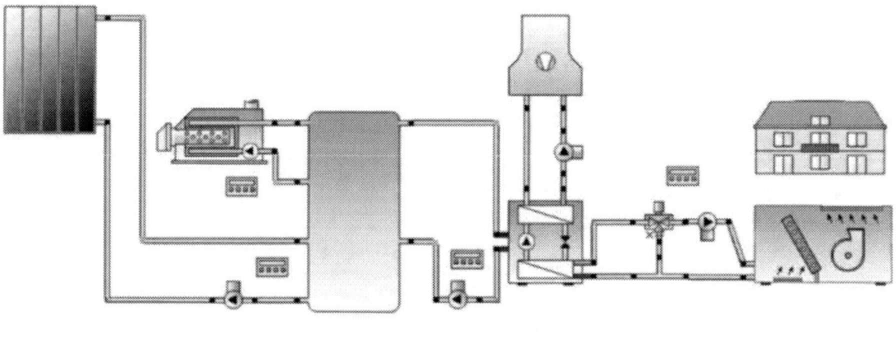

Fig. 3. Absorption chiller as a cooling technology and solar collectors as the primary heat generator (generated from Polysun simulation software)

enough). The main system components are: an absorption chiller (300 kW cooling capacity), a 1600 flat-plate premium solar collector, a gas boiler for auxiliary heat generation (capacity 200 kW), a storage tank (400,000 l) and a cooling tower for the condensation of working fluid in the chiller using ambient air.

The main aim of this system is to show the possibility of utilizing solar thermal energy in the cooling system based on the absorption cooling machine. In the system under discussion, standard flat/plate collectors can be used because the temperature in the chiller generator is not very high. This system was not optimized as it was only used to provide an example of how this particular technology might be used. There is room for improvement by using different types of the collector (vacuum-tube or concentrated), different solar fractions, different storage systems (tank-in-tank storage system, borehole storage or pit thermal energy storage) and different system configurations. The advantage of this system is the same as the photovoltaic system, in that the cooling demand occurs simultaneously with solar energy supply. Therefore, there is a possibility of using a smaller boiler, and only for the periods during which the capacity of the solar collector field is not sufficient. It is important to note that for any further work with this variant, attention must be paid to the stagnation in the solar collector field, and the necessary means for preventing such stagnation must be provided in the design project.

3.3 Economic Feasibility Analysis

The profitability tool embedded in the Polysun software as well as excel-calculator were used for these calculations. There were almost no economic data available from the site, therefore assumptions have to be met in order to calculate the profitability of the projects. With more accurate prices available, the feasibility of the project can be recalculated using the Polysun profitability calculations or excel calculator.

In order to compare different systems, the following key-figures have been chosen: investment costs, net-present value, annuity, renewable energy fraction and CO2 savings. The results of the comparison is shown in Table 1. The capital recovery factor (CRF) was taken 0.0546 for both systems i.e., by assuming that the real discount rate of 5% and number of annuities received of 40 year [1].

Table 1. Key figures comparison of solar cooling systems

Key Figures	PV Powered Cooling System	Solar Thermal Cooling System
Investment Costs [USD]	14,759,567	783,996
NPV	-2,987,924	-434,986
Annuity	163,068	23,740
Renewable Energy Fraction [%]	100	100
CO2 Savings [kg]/Year	271,390	487,498

It obviously seen that the investment cost for solar PV powered cooling system almost double of the solar thermal cooling system. This system has been designed based on the assumption that the available air condition system remains the same and operates well. However, it is worth to mention, that the installed capacity of the current system corresponds to the required capacity calculated with the rule of thumb for this type of buildings in Indonesia, and yet the set temperature is not met, and premises are heating up to 35 degrees during the day. Therefore, the technical evaluation and maintenance, possibly replacement with the new one have to be considered. In this case, the costs of maintenance, repairs or replacement have to be added.

The lower annuity was found for the solar thermal cooling system, which operates 100% on the solar thermal energy means there is no need to pay for electricity/gas consumption. Moreover, this system has the high rate of CO2 savings of 487,498 [kg] annually. The biggest drawback of the proposed system is the area needed to install the solar collector field. The negative NPV shows that the earnings generated from the investment did not exceed the anticipated costs. However, NPV does not consider the use-value of the air conditioning system for the library in the cash-flows. The advantage of increasing the quality of service provided by the library cannot be measured as a direct financial return. Indonesia is now striving to reform the subsidy policy on the electricity market in order to reduce the level of using coal for electricity generation and thus the level of GHG emissions. The removal of subsidies gives an opportunity to invest into the clean energy sources.

4 Conclusions

The simulation and economic analysis of solar cooling for building in tropical climates in Surabaya, Indonesia has been carried out. Two different possibilities of solar cooling technologies are simulated using Polysun software each for: (i) photovoltaic (PV) powered cooling system, and (ii) solar thermal absorption cooling system. For the PV powered option, about 376 kWp of PV panels required to full fill energy for cooling load. While for solar thermal absorption cooling system option, the main system components are: an absorption chiller (300 kW cooling capacity), a 1600 m2 flat-plate premium solar collector, a gas boiler for auxiliary heat generation (capacity 200 kW), a storage tank (400,000 l) and a cooling tower for the condensation of working fluid in the chiller using ambient air. From economic aspect, the investment cost for solar PV powered cooling system almost double of solar thermal cooling system, means that solar thermal absorption cooling system is more feasible than PV. This is due to the good agreement between solar radiation, efficiency, and cooling demand. Besides, this systems allows the maximum CO2 emissions savings.

References

1. Hanna Satnikova (2017), Solar cooling system for the university library in Surabaya, Indonesia. semester Project Report, ZHAW School of Engineering.
2. Casals, X. G. (2006). Solar absorption cooling in Spain, Perspectives and outcomes from the simulation of recent installations. *Renewable Energy*, 1371-1389
3. Ali H. , A. H., Noeres, P., & Pollerberg, C. (2008). Performance assessment of an integrated free cooling and solar powered single-effect lithium bromide-water absorption chiller. *Solar Energy*, 1021-1030.
4. Mateus, T., & Oliveira, A. C. (2009). Energy and economic analysis of an integrated solar absorption cooling and heating system in different building types and climates. *Applied Energy*, S. 949-957.
5. https://www.google.com/maps
6. Nationalelektronik.com (2017), Cara menghitung kebutuhan pemakaian PK AC sesuai ruangan. Access 19-02-2017. https://www.nationalelektronik.com/2017/01/cara-menghitung-kebutuhan-pemakaian-pk-ac-sesuai-ruangan/
7. Weather and Climate. (19. 02 2017). Von https://weather-and-climate.com/average-monthly-Rainfall-Temperature-Sunshine, Surabaya,Indonesia.

Effect of butanol on fuel consumption and smoke emission of direct injection diesel engine fueled by jatropha oil and diesel fuel blends with cold EGR system

Syarifudin[1,*], *Syaiful*[2] , and *Eflita* Yohana[2]

[1]EDP Sciences, Editorial Department, 91944 Les Ulis Cedex A, France
[1,2]Mechanical Engineering Department, Diponegoro University, 50275 Semarang, Indonesia

Abstract. Diesel engines are widely used in industry, automotive, power generation due to better reliability and higher efficiency. However, diesel engines produce high smoke emissions. The main problem of diesel engine is actually the use of fossil fuels as a source of energy whose availability is diminishing. Therefore alternative fuels for diesel fuels such as jatropha and butanol are needed to reduce dependence on fossil fuels. In this study, the effect of butanol usage on fuel consumption and smoke emissions of direct injection diesel engine fueled by jatropha oil and diesel fuel with cold EGR system was investigated. The percentage of butanol was in the range of 5 to 15%, jatropha oil was in the range of 10 to 30% and the balance was diesel fuel. Cold EGR was varied through valve openings from 0 to 100% with 25% intervals. The experimental data shows that the BSFC value increases with increasing percentage of butanol. In addition, the use of EGR results in a higher increase of BSFC than that without EGR. While the addition of butanol into a blend of jatropha oil and diesel fuel causes a decrease in smoke emissions. The results also informed that the use of EGR in the same fuel blend led to increased smoke emissions.

1 Introduction

Diesel engines are widely used due to higher fuel efficiency and generate greater power and torque compared to gasoline engines [1]. In addition, the diesel engine does not require spark plugs because the combustion system uses compression-ignition. This system allows the achievement of high initial pressure before the combustion process occurs [2]. Diesel fuel has a chemical structure made up of atoms and hydrogen with a straight chain structure. This structure is perfect because of better ignition quality so as to support the perfect combustion process [3]. The widespread use of diesel fuels, resulting in diminishing petroleum reserves and increasing gas emissions. Jatropha is a fuel derived from non-food ingredients studied as an environmentally friendly alternative fuel [4]. According to Jinlin Xue et al. [5], blending jatropha with diesel fuel can reduce fears of oil fuel depletion

* Corresponding author: masudinsyarif88@email.com

without sacrificing power and economic value of diesel engines, but the use of jatropha as a blend material results in higher carbon emissions compared to diesel [6].

Syaiful et al [7] in his research observed that by adding liquor alcohol (methanol) to jatropha and diesel fuel blends can reduce soot emissions, and this decrease increases with increasing percentage of alcohol in the blended. According to Imtenan S et al. [8], blending butanol on the use of jatropha-fueled diesel engines, resulting in lower carbon emissions compared to diesel fuels. In the study Fayad MA et al. [9], the active role of oxygen compounds contained in butanol as a blend of diesel fuel has an effect on increasing the oxidation concentration thereby suppressing the formation of soot. In addition, the high rate of heat propagation can decrease the size of Particulate Matter (PM). This is evidenced by Chen Guisheng et al. [10], by experimenting the addition of butanol by 20% (B20) and 30% (B30) on diesel fuel. The effect of the addition of B20 blended is slightly better than B30 in decreasing soot formation due to longer ignition time. Therefore, it can be concluded that the use of butanol and jatropha as a blended of diesel fuel with the EGR system has a positive impact on diesel engines, so research needs to be done specifically on specific fuel consumption and the resulting carbon black emissions.

2 Research Methods

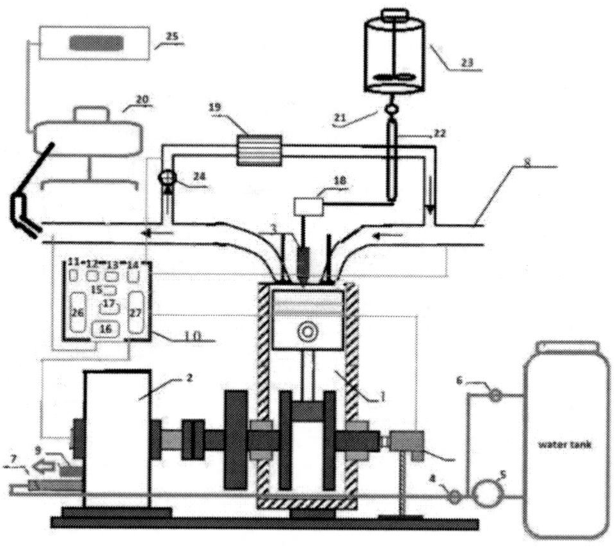

Fig. 1. Schematic of experiment.

Isuzu 4JB1 engine is released with a brake-water dynamometer. The printing press uses a digital tachometer with a proximity sensor. A measuring cylinder and stopwatch to measure the flow rate of fuel. The exhaust temperature was monitored by a digital thermometer, while the inlet flow and the EGR flow rate were measured by using the orifice. An experimental set up can be seen in Figure 1. The test was done at a constant engine speed of 2500 rpm, by varying the EGR band by 0%, 25%, 50%, 75% and 100%. In this condition, the EGR input is cooled by cooling the EGR channel to be inserted into the cylinder (cold EGR). The opening variations of 25%, 50%, 75% and 100% opening are done in any EGR deadlock situation. Variations are made for each fuel blended. Such a fuel in Table 1.

Table 1. Properties of alkohol and diesel.

No	Propertis	Diesel	Jatropha	Butanol
1	Viscosity (at 40°C (mPa.s))	2,0-5,0	3,23	2,63
2	Water content (%v)	0,05	3,16	34,22
3	Cetane number	48	41,8	17
4	Heating value (MJ/kg)	45,21	37,97	33,21
5	Flash point (°C)	60	198	35
6	Oxygen Content (%)	-	10,9	21,69

3 Result and Discussion

3.1 Data and Analysis of Fuel Consumption Test Result (BSFC)

Figures 2 (a) and 2 (b) illustrate the effects of fuel variations on BSFC with EGR applications and without EGR. The value of BSFC decreases as the percentage of load increases. This decrease is due to an increase in cylinder wall temperature which decreases the ignition delay. So that burning increases and fuel consumption decreases [11]. The use of blended fuel produces a higher BSFC value than pure diesel. Increased percentage of jatropha and butanol cause increasing BSFC value. This increase is due to the high concentration of oxygen in butanol [12]. The increase in the value of BSFC due to the percentage of butanol also occurred in the study of Ibrahim Amr [13], and Hua Tan Yie et al. [1]. Increasing jatropha content in blended fuel also resulted in increased BSFC value. This is due to the low value of heat contained in jatropha [14]. In addition, the opening of the EGR valve also causes the value of BSFC to be increased compared without EGR. The increase of BSFC value is due to dilution effect by decreasing the concentration of oxygen into the cylinder. As observed by Donghui Qi et al. [15].

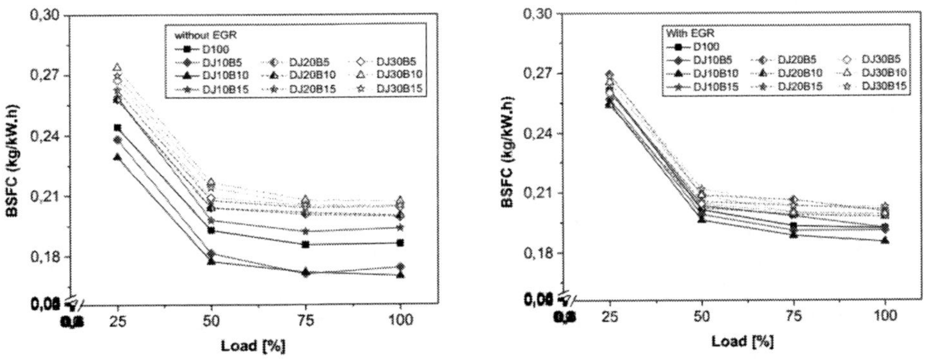

(a) (b)

Fig. 2. Comparison of fuel consumption for various butanol concentration for different engine loads without EGR system (a) and with EGR system (b)

Based on the above experiments, the use of blended fuels generally produces a higher BSFC value than the D100 except blended fuel with a jatropha content of 10% lower than D100. This is due to an increase in power value due to the high concentration of oxygen

 https://doi.org/10.1051/shsconf/20184902010

concentration in butanol which is more dominant than EGR. Testing with EGR on average yields a higher BSFC score than testing without EGR. In testing without EGR, the highest increase of BSFC value occurs when using DJ30B10 fuel at 50% load which is 12.33% and the highest BSFC value occurs at 0.273 kg/kW.jam fuel. While on the test with EGR application, the highest increase of BSFC occurred on DJ30B5 fuel at full load of 5.61% and the highest BSFC value occurred on DJ30B15 fuel at the low load of 0.268 kg/kW.jam.

3.2 Data and Analysis of Emissions of Soot (Opacity)

Figures 4 (a) and 4 (b) are a presentation of the effect of fuel variations on opacity with EGR applications and without EGR. Smoke opacity increases as a percentage of load increases. This is due to the reduced AFR at high loads where the amount of fuel injected is more into the combustion chamber so that much of the fuel is not burned and wasted with the exhaust gases [5]. But along with the increase in the percentage of butanol in blended fuels the smoke opacity decreases. This is evident from the diminution of particulate matter (PM) dimension in soot due to high oxygen content in blended fuel [9]. High oxygen content in alcohol (butanol) encourages the occurrence of carbon oxidation so smoke opacity becomes decreased [16]. The decrease of smoke opacity due to oxygen concentration also occurred in Huang Haozhong et al. [17], and Zerrakki M Is et al. [12]. In addition, the jatropha content in blended fuel also resulted in a decrease in smoke opacity [18]. This is due to the low aromatic compound and sulfur content in jatropha and the oxygen content present in jatropha causing a decrease in the value of smoke opacity [5].

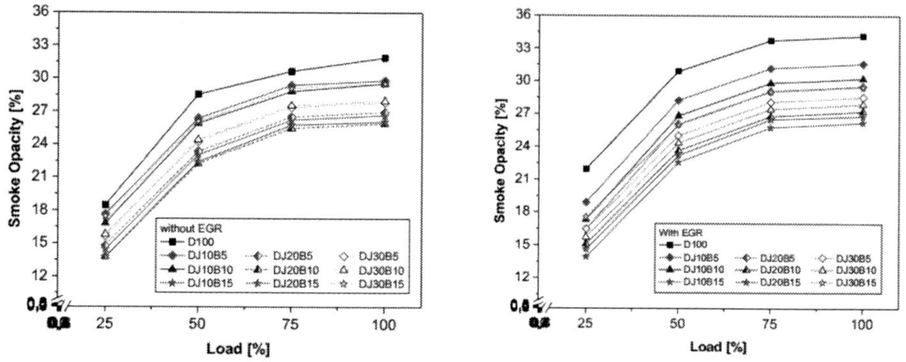

(a) (b)

Fig. 3. Comparison of Soot emissions (opacity) for various butanol concentration for different engine loads without EGR system (a) and with EGR system (b)

Based on the above test images, the use of blended butanol and jatropha fuels in diesel fuel with EGR applications resulted in a decrease in smoke opacity but still higher than without EGR. This is due to decreased carbon oxidation due to the EGR application which reduces the supply of oxygen in the combustion chamber (Syaiful et al, 2013). In the without EGR test, the highest smoke opacity drop occurred when using DJ20B15 fuel at low loading of 25.24% and the highest smoke opacity value occurred at D100 fuel of 31.88% at full load. While using blend fueled, the highest smoke opacity value occurs on the DJ10B5 fuel of 29.81% at full load. In the with EGR test, the highest smoke opacity drop occurred when using DJ20B10 fuel at a low loading of 33.46% and the highest smoke opacity value occurred at D100 fuel of 34.16% at full load. While using a blended fuel, the highest smoke opacity value occurs on the DJ10B5 fuel of 31.61% at full load

4 Conclusion

The following are the main conclusions drawn from research on specific fuel consumption and soot emissions from direct injection diesel engines with blended diesel fuel, jatropha, and butanol cold EGR:

1. Butanol and jatropha concentrations in diesel fuel cause an increase in the value of BSFC. The low heating value and high oxygen content is the cause of the increase in the value of BSFC. In addition, the EGR Application also leads to an increase in the value of BSFC.
2. Experiments with EGR result in an increase in the value of BSFC with a higher average than without EGR. The highest increase of BSFC value when testing without EGR occurs on DJ30B10 fuel at 50% load which is 12.33% and the highest BSFC value occurs at 0.273 kg/kW.jam fuel.
3. While on the test with the EGR application, the highest increase of BSFC occurred on DJ30B5 fuel at full load of 5.61% and the highest BSFC value occurred on DJ30B15 fuel at a low load of 0.268 kg/kW.jam.
4. Smoke opacity increases as a percentage of load increases. The use of butanol and Jatropha causes the smoke opacity value to be lower than D100. This is because the oxygen content in butanol and jatropha triggers the occurrence of carbon oxidation.
5. Testing of variation of fuel with EGR application also resulted in a decrease of smoke opacity value but still higher than without EGR. In the with EGR test, the highest smoke opacity drop occurred when using DJ20B15 fuel at low loading of 25.24% and the highest smoke opacity value occurred at D100 fuel of 31.88% at full load. While using a blended fuel, the highest smoke opacity value occurs on the DJ10B5 fuel of 29.81% at full load.
6. In the EGR test, the highest smoke opacity drop occurred when using DJ20B10 fuel when the low loading was 33.46% and the highest smoke opacity value occurred at D100 fuel of 34.16% at full load. While using a blended fuel, the highest smoke opacity value occurs on the DJ10B5 fuel of 31.61% at full load.

References

1. Yie Hua Tan, Mohammad Omar Abdullah, Cirilo Nolasco, NurSyuhada Ahmad Fauzi, Georgie Wong Abdullah. Elsevier. Energy Converion and Management, **132**, 54-64 (2017)
2. Heywood, John B.L. New York. Mc. Graw-Hill. Inc. (1988)
3. A. Amin, A. GadallaH, A.K. El Morsi, N.N. El-Ibiari, G.I. El-Diwani. Egyptian Journal of Petroleum. **25**, 509-514 (2015)
4. Chandra Pandey, V., Kripal Singh, Shankar Singh, J., Kumar, A., Singh, Bajrang., Singh, R. *Renewable and Sustainable Energy Reviews*, **16**, 2870-2883 (2012)
5. Xue, Jinlin., Grift, Tony., Hansen, A., 2011, *Renewable And Sustainable Energy Reviews*, **91**, 1761-1767 (2011)
6. Senthilkumar, R., Ramados, K., Manimaran, R. *Global Journal of Researches In Engineering Mechanical and Mechanics Engineering*, **13**, 1 (2013)
7. Syaiful, and Sugeng, 2017, *Nova*, **51**, 53-94 (2017)
8. S Imtenan, H.H. Masjuki, M. Varman, I.M. Rizwanul Fattah, H. Sajjad, M.I. Arbab. *Energy Converion and Management,* **94**, 84-94 (2015)
9. M.A. Fayad, A. Tsolakis, D. Fernández-Rodríguez, J.M. Herreros, F.J. Martos, M. Lapuerta. *Elsevier. Applied Energy,* **190**, 490-500 (2017)

 https://doi.org/10.1051/shsconf/20184902010

10. Guisheng Chen, Yinggang Shen, Quanchang Zhang, Mingfa Yao, Zunqing Zheng, Haifeng Liu, 2013. *Elsevier. Energy*, **54**, 333-342 (2013)
11. Dwivedi, G., Jain Siddarth, Sharma, MP. *Renewable and Sustainable Energy Review*, **15**, 4633-4641 (2011)
12. M. ZerrakiIsik, Hasan Bayindir, BahattinIscan, Dr. Huseyin Aydin. *Journal of the Energy Institute"*. **90**, 174-184 (2017)
13. Ibrahim Amr. *Applied Thermal Engineering*. **103**, 651-659 (2016)
14. Karabektas, M., Ergen, G., Hosoz, M. *Fuel*, **04**, 36-39 (2011)
15. Qi, Donghui., Leick, M., Liu, Y., Chia-Fon. *Fuel*, **90**, 1884-1891 (2011)
16. Syaiful, Stefan, M., Bae, M.W., Kazuo, T. *The Seventh international Symposium on Mechanics, Aerospace and Informatics Engineering (ISMAI)*, Japan, **07**, 147-151 (2013)
17. Huang HaozhongQingsheng Liu, WenwenTeng, Qingxin Wang, *Energy Procedia*, **105**, 914-920 (2017)
18. Gomma, M., Alimin, A.J., Kamarudin, K.A. *World Academy of Science, Engineering and Technology*, **38**, 522-527 (2010)

Analysis of the cause of the defect packaging of capsule products using six sigma: A case study (PT SM)

Finsaria Fidiyanti[1,*], and *Novie* Susanto[1]

[1]Department of Industrial Engineering, Diponegoro University, Jl. Prof. Soedarto, S.H., Tembalang, Semarang, Indonesia 50275

Abstract. This research aims to manage the amount of defective packaging of capsule products in the department of capsule product at PT SM. It is important because the number of defective packaging of capsule products is high and exceeds the limits of maximum, i.e. 5% of the total production. In addition, the defective packaging of capsule products cannot be reworked and will cause losses in terms of time and finances. This research uses the Six Sigma method to find out the capabilities of the production process from the defective packaging of capsule products. Application of the method of Six Sigma through DMAIC process to eliminate the number of defective products and Fishbone diagram is used to identify the cause of the defect and provide recommendations on improvements to the production process. The data for this research is the amount of production, defective packaging of capsule products and interviews with the department of quality control, department of capsule product and operator. Data processing using Six Sigma methods showed that the value of sigma is 3.41σ with the number of DPMO is 27,951.202. Using Fishbone diagram, there are 9 causes of the defective packaging of capsule products.

1 Introduction

The industrialized world has grown very rapidly. This is due to the needs of the society of a product improved in terms of quality as well as quantity supported by rapid advances in technology. With a large number of those needs, new industries have emerged that resulted in the onset of competition in the industrialized world, including the herbal industry.

According to the Ministry of the industry of the Republic of Indonesia, the national herbal medicine industry in the year 2017 experienced 10% growth over the previous year. This is due to the more various types of herbal medicine and the use of modern technology so that the machine is able to meet the needs of consumers who are increasingly diverse. In order to be able to survive in the competition of the industry, the company should improve the quality of products.

* Corresponding author: finsafidi@gmail.com

PT SM is a company engaged in the field of herbal medicine processing since 1940 and has certification that is equivalent to pharmacy standardization. Although it has become one of the largest herbal medicine company in Indonesia, PT SM has problems such as the large number of quality defects. A large number of defects that arise can affect productivity and losses both in terms of financial and time.

From the observations, the number of defects occurred in the department of capsule product section packaging. The number of defective packaging of capsule products exceeded the maximum i.e. 5% of the total production. In addition, the packaging product cannot be reworked so that it will cause losses in terms of time and finances. A large number of defective packaging of capsule products caused by the quality control not yet implemented properly. The company just did on logging the number and types of defects without identifying the defect causes even further. Therefore, it is necessary to improve quality control by minimizing the number of defective packaging of capsule products.

Thus, on the basis of the problem which is owned by the PT SM, this study aims to manage the process of quality control of packaging of capsule products by identifying the types and causes of disability. In addition, this research will also provide recommendations for improvements to manage the amount of defective packaging of capsule products.

This research uses the method of Six Sigma through DMAIC process (Define, Measure, Analyze, Improve and Control) and Fishbone Diagram. Six Sigma method is used to find process capability [1] on the department of capsule product section packaging, while Diagram Fishbone used to find out the causes of defects in the packaging of capsule products. The method of Six Sigma through DMAIC process has been used in previous studies. One of those researches is the research of Prabu Kolochi, Makesh, J K Naveen Raj, S.R. Devadasan & r. Murugesh in 2013 that use methods of Six Sigma through DMAIC process to find out the casing stage on the production of small and medium pumps [2]. This research was undertaken with the experimental design in the form of observations to obtained values of sigma 3.90σ. Other research is Kumar et al (2008) use the same method to improve customer [3] and Rohini and Mallikarjun (2011) that uses the DMAIC process in the healthcare industry in India [4]. It is illustrated that the method of Six Sigma through DMAIC process has a very broad implementation in various fields. Six Sigma method is performed to find out the sigma level of the production process of packaging of capsule products through the DMAIC process for planning quality control continuously [5] in order to achieve zero defects.

2 Literature Review

2.1 Six Sigma

According to Gasperz, Six Sigma is a quality improvement vision towards the target of 3.4 failures per one million opportunities for every transaction of goods and services products [6]. The goal is to increase the efficiency of business processes by means of emphasizing the removal of errors, removing waste and minimize the rework of defect products in order to increase the customer satisfaction so that the company's value has also increased.

2.2 DMAIC Process

There are five stages of implementing the Six Sigma method, that is Define – Measure – Analyze – Improve – Control (DMAIC) which is a recurrent phase [7].

Define
Is the process of identification and determination of priority issues for improving quality of goal setting. This stage consists of problem identification, SIPOC, and CTQ.
Measure
Is the process of identification of product defects mapping, performance measurement process to set goals increase. This stage uses a control chart for identification of any condition of uncontrolled in the process and the calculation of the DPMO and sigma levels.
Analyze
The process of analyzing and reviewing the data collected to determine the priority list of the source variation. The instrument used to determine the root cause of the problem is the fishbone diagram.
Improvement
Stage of the improve process by removing the cause of the defect. This stage is done by making changes to achieve increased quality towards zero defects.
Control
Stage to ensure the control is done well underway and continuous, also prevent the occurrence of the failure. This stage is done with documentation and dissemination of successful procedures as the standard guidelines.

3 Research Methodology

The data processing on this research using a quantitative approach, so that needs to calculate of adequacy test, uniformity test and normality test of the data obtained. Adequacy test is used to ensure that data that has been collected is quite objectively. This test is performed by comparing N ' and N where N' was obtained from calculation. If $N' \leq N$, the data is considered sufficient. Uniformity test is used to ensure the uniformity of the data collected comes from the same system. This test is performed by calculating the upper limit and lower limit. If the data contained in both the limit, the data is considered uniform. Normality test used to assess the distribution of the data on a group of data. This test using SPPS software. If $KS < 0.235$ or $P\text{-value} > 0.05$, the data is considered to be normally distributed. If the data is sufficient, uniform, and normal, then performed the data processing by using the method of Six Sigma DMAIC process. On DMAIC process, using SIPOC diagram on stage define, u control chart on stage measure and fishbone diagram on analyzing phases.

4 Result and Discussion

DMAIC Process

4.1 Define

4.1.1 Problem Identification

The number of the defective packaging of capsule products exceeded the maximum i.e. 5% of the total production and cannot be reworked so that it will cause losses in terms of time and finances.

4.1.2 Identification of Key Process (SIPOC)

Figure 1 shows the SIPOC diagram of the packaging process of capsule product.

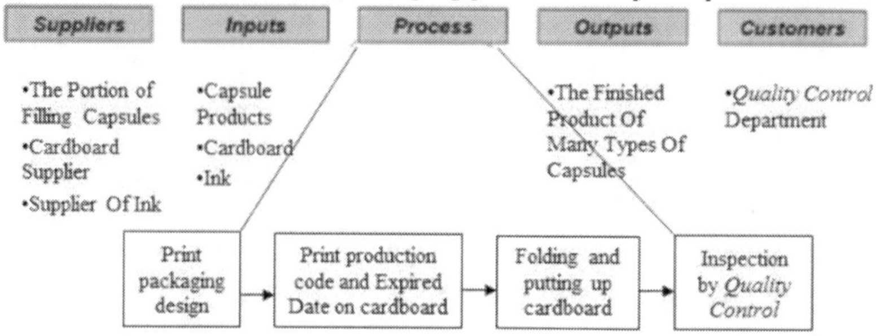

Fig. 1. SIPOC Diagram of the Packaging Process of Capsule Product

4.1.3 Identification of Critical to Quality (CTQ)

CTQ (Critical to Quality) is a key quality characteristic and it is directly related to the specific needs of the customer. From the results of the interview with the department of quality control, department of capsule product and operator, gained 3 types of defects that often occur, they are:

1) Image Position Won't Fit (image not printed in the right position)
2) Shaded Results (there is a shadow on the image)
3) Color does not match (color image does not suitable to the standard)

4.2 Measure

4.2.1 u Control Chart

Iteration 0

Figure 2 illustrates the results of the u control chart in iteration 0, there are 3 data that is out of control and need for elimination.

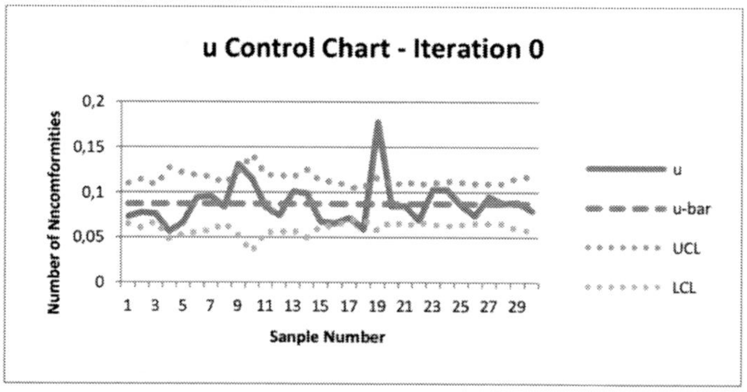

Fig. 2. u Control Chart – Iteration 0

Iteration 1

Figure 3 illustrates the results of the u control chart in iteration 1, all data in control. This shows that the defect number is controlled statistically

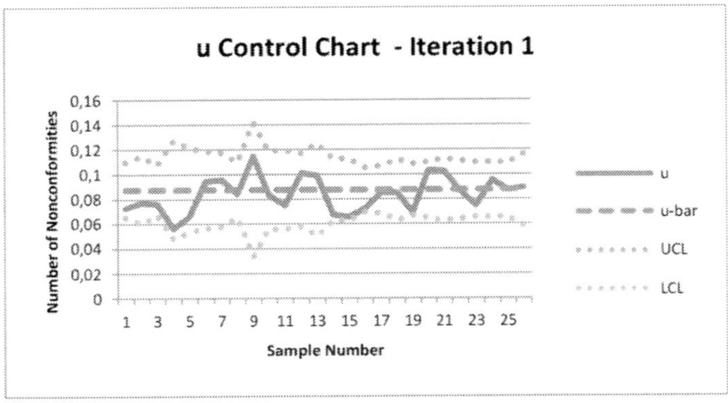

Fig. 3. u Control Chart – Iteration 1

4.2.2 Measurement of the level of DPMO and Sigma Levels

Determination of Value of Sigma Period

TOP (*Total Opportunities)* = Total product produced x Number of CTQ
= 1542 x 3 = 4626

DPO (*Defect Per Opportunities*) = $\frac{D}{TOP} = \frac{112}{4626}$ = 0,024 probability of defect for one occurence

DPMO = DPO x 10^6 = 0,024 x 10^6 = 24211

Sigma Period = NORMSINV((1000000-DPMO)/1000000)+1,5 = 3,47

Determination of Value of Sigma Process

TOP (*Total Opportunities)* = 37160 x 3 = 111.480

DPO = $\frac{3116}{111.480}$ = 0,028 probability of defect for one occurence

DPMO = 0,028 x 10^6 = 27951,202

Sigma Period = NORMSINV((1000000-DPMO)/1000000)+1,5 = 3,41

Figure 4 and Figure 5 shows the diagram of comparison of sigma value and diagram of comparison of DMPO value.

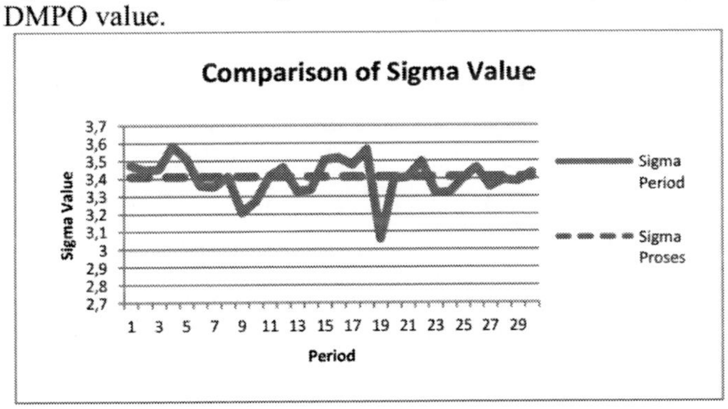

Fig. 4. Diagram of Comparison of Sigma Value

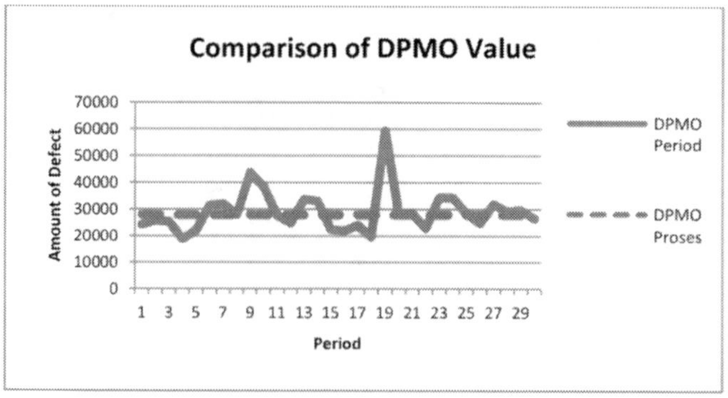

Fig. 5. Diagram of Comparison of DPMO Value

Sigma level from the calculation is 3.41 σ.

4.3 Analyze

4.3.1 Increasing Sigma Value

It is assumed, every period (30 days) the company should be able to increase the value of sigma of 0.1 to achieve the value of 4σ.

Increasing sigma (%) = $\frac{Sigma\ target - Sigma\ Baseline}{Sigma\ Target} x\ 100\%$ = $\frac{4-3{,}41}{4} x100$= 14,75%

DPMO to be achieved is 6209.72, then:

Decreasing DPMO (%) = $\frac{DPMO\ Baseline - DPMO\ Target}{DPMO\ Baseline} x\ 100\%$ = $\frac{27951{,}202 - 6209{,}72}{27951{,}202} x100\%$
= 77,78%

Table 1. Recapitulation of Sigma and DPMO Value.

Period	Number of Units	*Sigma value*	CTQ	DPMO	DPO	TOP	Number of Defects
1	37160	3.41	3	27951.202	0.02795	126716	3542
2	37160	3.508	3	24328	0.02433	130370	3172
3	37160	3.607	3	20704	0.02070	134024	2775
4	37160	3.705	3	17080	0.01708	137678	2352
5	37160	3.803	3	13457	0.01346	141332	1902
6	37160	3.902	3	9833	0.00983	144986	1426
7	37160	4.000	3	6209.72	0.00621	148640	923

From the table, it can be targeted within ± 7 months to reach 4σ.

4.3.2 Fishbone Diagram

Figure 6 shows root and cause of the defective packaging of capsule products using fishbone diagram.

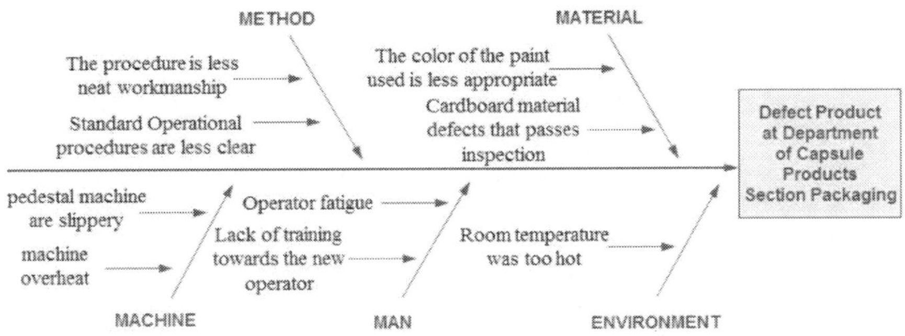

Fig. 6. Fishbone Diagram of the Defective Packaging of Capsule Products

4.4 Improvement

Some of the proposed improvements that can be made:

1) Make of Standard Operational Procedure clearly and detail for each process to minimize defect products and reduce the occurrence of waste material.
2) Conducts an inspection of the color of the paint prior to the printing process.
3) Tighten the inspection on quality control department in order for the defect cardboard material that is obtained from suppliers can not enter the production process.
4) Replace the base printing engine with a textured base so it can not change the position of the cardboard when the process of printing.
5) Checked the temperature of the machine periodically to avoid overheat engine when the printing process.
6) Additional of rest period into 45-60 minutes to reduce the level of tiredness on workers.
7) Conduct training for the operator to equate the capabilities of operators as well as provide knowledge in accordance with standard procedure.
8) Provides additional air conditioner in order to make the room cooler and accelerated the drying process of results printing.

5 Conclusion

From the research, the problem is the number of the defective packaging of capsule products exceeded the maximum i.e. 5% of the total production and cannot be reworked so that it will cause losses in terms of time and finances. To improve the quality of the product, quality control and quality improvement are required. Using Six Sigma, it is found that the actual sigma value is 3.41σ. It needs to improve sigma for improvement of product quality. Sigma improvement needs +/– 7 months to reach 4σ. Some problems that happened are image position won't fit, shaded result and color do not match. Some of the recommendations that can be given are make of Standard Operational Procedure clearly and detail, tighten the inspection of materials, replacement of the base printing machine with a textured base, checked the temperature of the engine periodically to avoid engine overheat, addition of a rest period into 45-60 minutes, conduct training for operators and provide additional air conditioner.

References

1. D. A. Desai, *Int. J. Productivity and Quality Managem.*, **9**, 258 (2012)

2. P. Kolochi, J. Makesh, K. N. Raj, S.R. Devadasan, R. Murugesh, *Int. J. of Process Managem. and Benchmarking,* **3**, 386 (2013)
3. S. Kumar, M. Sosnoski, *Int. J. of Productivity and Performance Managem.*, **58**, 254 (2009)
4. R. Rohini, J. Mallikarjun, *Procedia – Social and Behavioral Sci.*, **25**, 273 (2011)
5. P. S. Pande, R. P. Neumann, R. R. Cavanagh, *The Six Sigma Way Bagaimana GE, Motorola & Perusahaan Terkenal Lainnya Mengasah Kinerja Mereka*. Yogyakarta, Indonesia: ANDI (2002)
6. V. Gasperz, *Total Quality Management*. Jakarta, Indonesia: Gramedia Pustaka Utama (2005)
7. Kementerian Perindustrian Republik Indonesia. *Industri Jamu Tumbuh 10 Persen dan Serap 15 Juta Orang*. Retrieved from http://www.kemenperin.go.id/artikel/17906/Industri-Jamu-Tumbuh-10-Persen-dan-Serap-15-Juta-Orang (2017)
8. S. Sumit, D. Kaushish, *Int. J. of Sci. and Res.*, **4**, 450 (2015)

Analysis of Scraps Cause of Noodle Production in PT. Indofood CBP Success Makmur Tbk. using Fault Tree Analysis Methods

Diana Puspita Sari[1,*], *Sela* Tri Parwati[1], *Dyah* Ika Rinawati[1] and *Purnawan* A. Wicaksono[1]
[1]Industrial Engineering Department, Faculty of Engineering, Diponegoro University
Jl. Prof. H. Soedarto, SH, Tembalang, Semarang, Indonesia

Abstract. Food industry can survive when they can improve the productivity to meets the customer needs. PT Indofood CBP Sukses Makmur Noodle Division is one of food industry in Indonesia. Based on the production data of PT Indofood CBP Sukses Makmur Noodle Division Cibitung plant 3, found that the number of scrap noodles is almost always above the company minimum limit. This condition causes the company losing a profit of about 1.3 million per day per line. Based on these problems, this study aims to find the root causes of scraps of noodles production, then provide the advice on prevention of high amount of scrap of noodle production. The method used in this research is Fault Tree Analysis (FTA) and Five Step Plan analysis. The result of FTA analysis shows the cause of the high scrap of noodle production is the number of blocks noodles that fall as it passes through the conveyor in the frying and packing process, and because the operator less observant in connects the spice etiquette, less thorough to set the speed of the autoloader engine, less careful in case of handling the noodles. Suggested solutions include the implementation of 'seiri' by sorting reject spice oil and spices; 'seiton' by placing rolls of etiquette in easily reach place by the operator; 'seiko' by discarding etiquette, spices, and spiced oils into their respective containers; 'seiketsu' which is the division of responsibility for cleanliness of each operator; and 'shitsuke' ie routine briefing based on schedule.
Keywords: Fault Tree Analysis (FTA), Five Step Plan, Scrap

1. Introduction

As one of the industrial companies, especially food sector companies in Indonesia, PT Indofood CBP Sukses Makmur Noodle Division should always improve all aspects of the

* Corresponding author: dpsari.01@gmail.com

company in order to improve the quality. Feigenbaum explains that quality is the overall characteristic of both products and services [1]. Juran in Kaban [2] describes the definition of quality as a package of features and characteristics of products that provide satisfaction for consumers, conformance to requirement that is something that has been standardized, conformity with the needs of the market, or the state of the product of the physical, nature, and function so that it can meet the wants and needs of consumers in accordance with the value that has been issued. The product is said to be of quality if the product has a suitable use for itself. Another definition says quality is the goods or services that can raise the user's status, others say goods or services that benefit the user (measure of utility and usefulness) [3]. The main purpose of quality control is to obtain assurance that the quality of the product or service is produced in accordance with predetermined quality standards by issuing economical or lowest cost [4]. Garvin (1995) suggests eight dimensions or critical categories of quality namely: performance, featured, reliability, conformance, durability, serviceability, aesthetics and perceived quality. A product will be declared qualified by the manufacturer, if the product is in accordance with its specifications [5].

PT Indofood CBP Sukses Makmur Noodle Division is the largest noodle division company in Cibitung. The production process involves making noodles from raw materials in the form of wheat flour to noodle blocks of noodles in the packaging of etiquette. The process of producing noodles produces scrap. Scrap is the result of the processing of noodles in the form of dough noodles, exsteam noodles, fractions of noodles, broken noodle blocks, and noodle blocks that are not in accordance with the standards so it is not feasible to proceed to the packaging and distribution process. From the various stages of this noodle making from raw materials to noodle blocks, of course there is activity/something that does not give added value, has no benefits, and also result in waste. Waste in the production process is grouped into 7, i.e.: excessive production, stock, damage, ineffective transport/transfer, unnecessary movements and waiting (Pramono, 2013).

Based on production data of PT Indofood CBP Sukses Makmur Noodle Division Plant 3, it is found that the number of scrap noodles is almost always above the minimum limit set by the company. Analysis of scrap production data of December 2016, obtained the value that every day production process produces the amount of scrap that is above the company standard limit (0.3), with an average percentage of 0.38. This causes the company losing a profit of about 1.3 million, or more than 1% of the profits received by the company per line per day. This scrap case becomes very detrimental to the company because the amount of scrap that exceeds the standard of PT Indofood and resulted in the company losses 2, 8 million in one month. Therefore it is necessary to do an analysis to determine the factors that cause a lot of scrap that arises.

The environmental and operation analysis of a process, to find solutions to existing problems can be explained by the Fault Tree Analysis (FTA) method. FTA is also called an error tree. Baig et al [6] explains that FTA is a very effective tool for assessing risk especially in complex systems. The advantages of the FTA is to begin solving the problem of the top event selected by the user as interest and the tree to be identified the root cause. The objectives of this research are to identify factors causing high number of normal noodle scrap at line 31 plant 3 PT Indofood CBP Sukses Makmur Cibitung, and to analyze the factors causing high number of scrap. From various factors that have been determined then done suggestions to PT Indofood as an effort to reduce the number of scrap that arises.

2. Research Methods

A preliminary study was conducted by collecting secondary data of scrap at line 31 plant 3 production of PT Indofood Cibitung. Based on these data, the amount of scrap production is almost always above the standard limit of the company. Then analyzed based on scrap production data of December 2016, the average percentage of scrap production amounted to 0.38 is above the average standard of 0.3 percent.

FTA diagramming begins by interviews and brainstorming with section manager, shift section manager, section and staff in the production and quality control. Based on the data then performed an analysis using the FTA method. Roberts and Vessely [7] explained FTA chart described the logical relationships of the underlying causes of the problem. This error tree is a graphical representation of parallel errors and combinations of errors as the cause of the errors that exist on it. FTA uses symbols to explain the type of event that causes the problem. The symbols used in the FTA are described in table 1. Gharahasanlou et al [8] explains that the FTA diagram uses the “AND” and "OR" gate as logical operators. At the "AND" gate the output event occurs only when the event occurs. While on the "OR" gate event output occurs if one event occurs.

Table 1. Meaning of FTA Symbols

Simbols	The Meaning
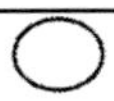	*Basic Event*, basic fault initiation that does not require further development
	*Conditionally Evently*, specific conditions that can be applied to various logic gates
	*Undevelopment Event*, event that cannot be developed anymore because the information is not available
	*External Event*, the expected event appears
	*AND gate*, the error arises because all input problems occur together
	*OR gate*, the error arises from one of the input problems that occur

The six basic steps often used in developing FTA [8] includes understanding of system configuration, build logic model, qualitative evaluation of the logical model, analysis of equipment failures and obtaining basic research data, quantitative evaluation of the logical model and proposed improvement.

The FTA making process begins with the definition of the problem and then proceeds to study the existing system and describes it in graphical form of the error tree. The steps taken in the manufacture of FTAs are: (1) Determining the top event of the problem, top event is the determination of the main problem in non-conformity that occur in a system, (2) Determination of limits used in FTA, learn the system by finding out Standard Operational Procedures (SOPs) or anything related to the company's mechanisms that maybe affect the error and (3) Designing FTA, FTA charting is made by reconsidering the various things that are considered as the cause of the problem with various information that has been obtained in the previous stage [7]:

Based on this FTA chart, the system logical relationship between the impacts and causes can be known. The main causes of problems can be found based on qualitative and quantitative analysis on the fault tree. This provides a reliable basis for determining how precautions are needed to prevent problems or to achieve goals. Hu [9] explains that the FTA is both a practical and comprehensive method for risk analysis and problem prevention, which is used in identification, prediction, evaluation and control. From the various events that are the main cause of the high number of scrap, then given solution suggestion using Five Step Plan method. This method aims to structuring and maintenance of work areas. Then this concept is called the Kaizen Concept [10], which consists of:

1. "Seiri" (Brief-Circle-Order). Seiri is a first step to run the 5S culture. Seiri has the intention to throw/sort/get rid of the goods, files that are not used again to the disposal.
2. "Seiton" (Neat-Arrange-Neatness). After sorting all items or files that are not used anymore, make sure everything must be placed according to the position set, so it is always ready to use when needed.
3. "Seiso" (Clean-Sweep-Clean). After being neat, the next step is cleaning the workplace, work space, equipment and work environment.
4. "Seiketsu" (Care-Uniform-Consolidation). This stage takes care of the three previously executed stages on a regular basis. This stage can also be called the stage of treatment, is the standardization and consistency of each individual to perform the previous stages.
5. "Shitsuke" (Diligent-Always-Discipline-Habituation). Maintenance of personal discipline includes a habit and maintenance of an already running 5S program.

3. Discussions

3.1 Designing of Fault Tree Analysis (FTA)

The first stage is done FTA analysis by identifying top event. Problems encountered in normal noodle production process PT Indofood CBP Sukses Makmur Tbk. Cibitung is the high number of scrap noodles production. The number of scraps of this noodle product is very high beyond the company standard limits. The problem of high amount of scrap is caused by various factors such as technical factor and human factor.

The second stage is determination of FTA limits. Limitations in this FTA are: normal noodle for AB type and under normal conditions. The third stage is creation of FTA. At this stage the FTA chart is made by reconsidering the various causes of scrap. From the FTA chart, various events are found at the bottom of the chart which is the basic event and undevelopment event as the root cause of scrap. The amount of scrap produced by normal noodle production at line 31 is triggered by two things, that is, because of the scrap caused at the frying stage and the unsuitable noodle block (at the packing stage). Of the two things that cause the occurrence of scrap, both occur simultaneously. Because if the frying event does not occur simultaneously with an unsuitable noodle block it will not cause a high number of scrap noodles.

The unsuitable event block of noodles is caused by 2 events, the noodle blocks of noodles affected by the end sealer event and noodle blocks of noodles that fall from the conveyor event do not occur simultaneously. This means that if one event of the noodle block is exposed to end sealer event and noodle blocks are not suitable event to occur individually then this will still cause a high amount of scrap. Event noodle blocks exposed end sealer caused by autoloader is jammed event and speed of autoloader event not match which do not coincide. From each event that happens separately then scrap can still cause high number of noodle scraps. Event noodle blocks fell from the conveyor caused by 3 events that occur separately. The three events are event of operator handling that are not good, the event of noodle blocks that hit the blower sensor, and the event of a gap in the conveyor connection that causes the noodle blocks to fall. The FTA chart of the causes of the high number of normal noodle production scraps in line 31 of PT Indofood's production plan 3 is illustrated in Figure 1. Based on the FTA analysis, table 2 shows the recapitulation of factors causing the occurrence of scrap.

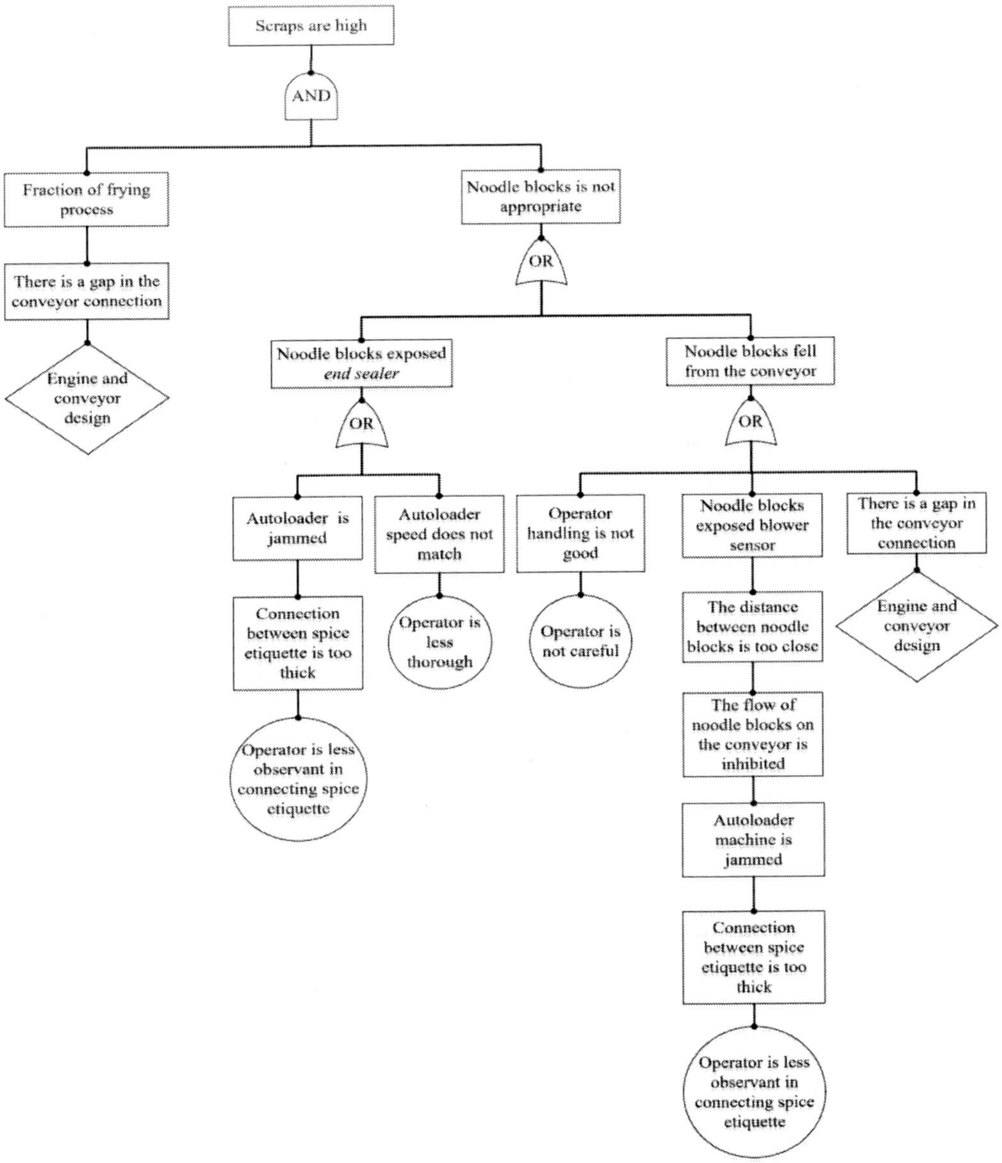

Figure 1. Fault Tree Analysis of Noodle Scrap Causes

3.2 Recommendation of Prevention using Five Step Plan

At this stage of giving recommendation to reduce the amount of scrap production. Based on the FTA chart, the cause of the number of scrap noodles at line 31 is preceded by the operator who is less observant, less thorough and less careful. Suggestions that can be given to the company, especially the production using the five step plan are:

1. Seiri (Sorting)
   The implementation of 'seiri' is sorting spice oil and spices that are rejected by the seasoning operator, then removing or disposing of spice oil or spices that reject it to the dump. So the seasoning oil and spices on the desk near the operator are spice oil and spices that do not leak (good). This will certainly increase the concentration of operators.
2. Seiton (Setup)
   The implementation of 'seiton' is done with the aim that the operator can find the goods easily, so the time becomes more effective. Implementation is by putting the reels of etiquettes in a corner line near the sealer, so the sealer operator is easy to take it.
3. Seiko (Hygiene)
   The implementation of 'seiko' can be done by removing the defective etiquette into the etiquette container, and disposing of damaged spices or spice oil or reject as soon as possible. It is aimed to avoid the accumulation of goods that are not used in the production process in the operator's work area, so the operator feels more comfortable. Hygiene supervision is required in the work area operator. Operators also have to get used to cleaning the workplace. In addition to maintaining cleanliness and production quality, this can also improve time efficiency and can reduce the chance of damaged machines.
4. Seiketsu (Care)
   The implementation of 'seiketsu' is done by dividing the responsibility of each operator on the cleanliness in their respective work area. From the management side, especially the shift section as the lower level management that always deal directly with the operator also must always provide control and supervision to the operator.
5. Shitsuke (Habituation)
   Implementation of 'shitsuke' can be done with the implementation of each briefing before shift begins by the related shiftsection. This will be a habituation for the operator to have a better atmosphere because the operator will prepare yourself early. Implementation of ISO training related to the cleanliness of the environment and the use of Personal Protective Equipment (PPE) with routine according to schedule.

Table 2 Recapitulation of Factors Causing the Incidence of Noodle Scrap

No	Cause	Description
1	Engine design and conveyor (on conveyor frying)	Noodles (in noodle blocks and crumb noodles form) fall from the conveyor connection after the frying process. The vibrating conveyor design has the purpose to drain the noodles that have just passed the qualifier. The connection between the qualifryer conveyor and the cooling conveyor which has a large gap and has a working mechanism to reverse the noodle block. in this case the design of the engine and conveyor becomes undevelopment event because there is no data that allows related to this matter.

No	Cause	Description
2	The operator is less observant in connecting spice etiquette	The thickness of the connection resulted in the autoloader engine jammed. Spice etiquette connections that are too thick also over time can make the machine damaged. Therefore, the need for handling of this basic event.
3	The operator is not thorough	This is related to the wrong set of autoloader engine speed, so it does not sync with the speed of the conveyor. If there is a wrong setting speed of the autoloader, it causes during the running, then the noodle blocks passing through this stage will be exposed to the end sealer. Therefore, the need for anticipation in order to avoid the basic event is happening.
4	Operator is not careful	This is related to operators handling that are less careful, resulting in many noodle blocks of noodles that fall to the floor (reject). Basic event is also associated with the mental atmosphere of workers. If the operator is hurried then lead to higher disparity. Therefore, the need for handling to minimize scrap generated by this basic event.
5	Operator is less observant in connecting spice etiquette	Rough connections resulted in the autoloader engine jammed, resulting in blocked noodle blocks. Therefore, the need for prevention of basic event is not the case.
6	Machine and conveyor design (packing conveyor)	Noodles (in the form of noodle blocks and crumb noodles) fall on the packing conveyor. However, unavailability of data that can be used causes this cause to be an undevelopment event

4. Conclusion

The high number of scrap is caused by engine and conveyor design (on frying and packing conveyor), operator less observant, less precise and less careful. Cause of high number of scrap are noodle blocks and crumbs of noodles falling from the conveyor joint after the frying process; the roughness of the connection resulted in the autoloader engine jammed, speeds of the autoloader engine and conveyor does not sync, operators handling that are less cautious, resulting noodle blocks of noodles that fall to the floor, the roughness of the connection resulted in the autoloader engine jammed, causing the blocks to be inhibited; and noodle blocks and crumbs noodles fall on the packing conveyor.

The preventive solution using the Five Step Plan include the implementation of 5S by sorting spice oil and spices rejected, then remove or remove the spice oil or spice that reject it to the dump (seiri), placing the reels of etiquettes in a corner near the sealer, so the operator can easily pick it up (seiton), removing the defective etiquette into the etiquette container, and disposing of damaged spices or spice oils or rejects as soon as possible (seiko), dividing the responsibilities of each operator over cleanliness in their respective work areas (seiketsu), and briefing before the shift begins by the related shift section and implementation of ISO cleanliness training and the use of Personal Protective Equipment (shitsuke).

References

1. Ariani, D. W. (2004). *Pengendalian Kualitas Statistik.* Yogyakarta: Andi Offset.
2. Kaban, R. (2014). Pengendalian Kualitas Kemasan Plastik Pouch Menggunakan Statistical Procces Control di PT Incasi Raya Padang. *Jurnal Optimasi Sistem Industri*, 13(1), 518-547.
3. Gasperz, V. (2001). *Metode Analisis untuk Peningkatan Kualitas*. Jakarta: Gramedia Pustaka Utama.
4. Assauri, S. (1998). *Manajemen Produksi dan Operasi*. Jakarta: FE Universitas Indonesia.
5. Juran, J. M. (1998). *Juran's Quality Control*. New York: McGraw-Hill.
6. Baig, A. A., Ruzli, R., & Buang, A. B. (2013). Reliability Analysis Using Fault Tree Analysis: A Review. *International Journal of Chemical Engineering and Application*, 4(3), 169-173.
7. Vessely, W. E., Goldberg, F. F., Roberts, N. H., & Haasl, D. F. (1981). *Fault Tree Handbook*. Washington DC: U.S. Government Printing Office.
8. Gharahasanlou, A. N., Mokhtarei, A., & Khodayarei, A. (2014). Case Study Fault Tree Analysis Failure Cause of Crushing Plant and Mixing Bed Hall at Khoy Cement Factory in Iran. Case Studies in Engineering Failure Analysis, 2(1), 33-38.
9. Hu, Y. (2016). Research on the Application of Fault Tree Analysis for Building Fire. *Proceedia Engineering*, 135, 524-530.
10. Suwondo, C. (2012). Penerapan Budaya Kerja Unggulan 5S (Seiri, Seiton, Seiso, Seiketsu dan Shitsuke) di Indonesia. *Jurnal Magister Manajemen*, 1(1), 29-48.

Working posture analysis of sweet whey powder handling at CV Cita Nasional warehouse using OVAKO Working Posture Analysis (OWAS)

Manik Mahachandra[1,*], *Heru* Prastawa[1], and *Dhialma Yonathan* Susilo[1]

[1]Industrial Engineering Department, Faculty of Technology, Universitas Diponegoro, Semarang, Central Java, Indonesia 50275

Abstract. In this 4.0 industrial era, some material handlings are still conducted manually by human. One example is in the warehouse area of CV Cita Nasional, a milk processing factory in Kopeng, Central Java. Workers are repetitively unloading a 25 kg sack of sweet whey powder from truck to warehouse area. Therefore, this study was carried out to investigate the musculoskeletal disorders risks on that workers manual handling activity. The OVAKO Working Posture Analysis System (OWAS) was chosen to be applied in the evaluation steps. For each activity in the unloading process, OWAS score category of 3 (improvements are necessary as soon as possible) were resulted on lifting, 2 (improvements are necessary in the long-term future) on carrying, and 3 on lowering the load. Consequently, a trolley was then designed to minimize the risks. The trolley was designed based on Indonesian anthropometry dimensions. Applying this proposed tool, lifting and lowering activities were then categorized in OWAS score of 2, whilst carrying was resulted in OWAS score of 1 (improvements are not necessary). This applied study was valuable in order to minimize the musculoskeletal disorders risks on such manual material handling activity, not only in the current workplace, but also in similar situation of industries.
Keywords: anthropometry, manual material handling, musculoskeletal risks, OVAKO Working Posture Analysis System (OWAS), tool design

1 Introduction

CV Cita Nasional is one of the milk processing factory, located in Kopeng, Salatiga, West Java, Indonesia. The milk products, branded "Susu Segar Nasional", are very famous in the community. One of the main ingredients in the milk processing is the sweet whey powder. This ingredient made of secondary product of processed cheese, which involved enzyme as the coagulation method to absorb water, generate emulsification, and result foams. Whey protein is also commonly added in the processed food to increase the nutrition. The sweet whey powder used in this factory is produced by Poland company, ordered regularly every

* Corresponding author: manik.mahachandra@gmail.com

six months, and stored in specific warehouse in this factory. At the warehouse, the company hired six workers to handle all the storages. From a preliminary study, we configured that the workers' activity in storage handling were mostly conducted manually, which theoretically will develop high risks of musculoskeletal disorders (MSDs). Such prolonged situation, if not well controlled, will lead into injury and accident at work.

A manual material handling of sweet whey powder starts at lifting a 25 kg sack from truck to warehouse area. The load is usually carried on the top of worker's head, while his hands maintain the load along the process. As stated in [1], a lifting products task is more ergonomically hazardous for MSDs than the pulling task. Musculoskeletal disorder is a pathology condition that affect the normal function of musculoskeletal system of human body, including nervous, tendon, muscle, and supporting structure like discus intervertebral [2]. The disorder occurs as a result of repetitive activity and/or awkward working posture. The symptoms are including pain, sore, or cramps while working, or after work hours. As [3] reported in their research, 47% warehouse operators were involved in lower back injury, while 31% had knee problem and 21% of them experienced neck injury. Moreover, these numbers were related to 200 days of absenteeism. However, musculoskeletal disorders not only affected the workers' physical condition, but also the company. High number of absenteeism, high turnover, low motivation, and high stress level has been reported along with workers' injuries occurrences [4][5].

To prevent further problems of CV Cita Nasional warehouse workers, this study aimed to minimize the musculoskeletal risk of injury. Body postures and movements while working were captured and analyzed. A working tool was then designed to minimize the injury risk.

2 Method

2.1 Participants

Six male workers were contributed in the study as participants. All of them performed at least one activity in lifting, carrying, and lowering the sack of powder. They were all familiar with the layout and handling procedure, so that no particular preparation was needed before the data collection.

2.2 OVAKO Working Posture Analysis

The main data in this study were the working postures while moving the sweet whey powder from trucks to warehouse area. These postures were analyzed using the OVAKO Working Posture Analysis System (OWAS). OWAS was developed in a steel industry company, Ovako Oy, in Finland in 1973 to define the workload in the overhauling of iron smelting ovens [6]. OWAS is a suitable tool for analyzing the musculoskeletal risk when the whole body parts are working simultaneously, just like moving the load in this current study.

OWAS verifies the safety level of the most common work postures for the back (four postures), arms (three postures), and legs (seven postures), as well as the weight of the load handled (three categories). The quantification of each body parts is shown in Table 1, so that the whole body posture is described by those four body parts using a four digit code. It was then classified into four action categories indicating the needs for ergonomic changes, as presented by Fig. 1.

Table 1. OWAS action level.

Action		Digit
Back	Straight	1
	Bent	2
	Twisted	3
	Bent and twisted	4
Arms	Both arms below shoulder level	1
	One arm at or above shoulder level	2
	Both arms at or above shoulder level	3
Legs	Sitting	1
	Standing on two straight legs	2
	Standing on one straight leg	3
	Standing or squatting on two bent legs	4
	Standing or squatting on one bent leg	5
	Kneeling	6
	Walking	7
Load	Less or equal to 10 Kg	1
	Greater than 10 Kg and less or equal to 20 Kg	2
	Greater than 20 Kg	3

Back	Arms	Legs																				
		1			2			3			4			5			6			7		
		Load			Load			Load			Load			Load			Load			Load		
		1	2	3	1	2	3	1	2	3	1	2	3	1	2	3	1	2	3	1	2	3
1	1	1	1	1	1	1	1	1	1	1	2	2	2	2	2	2	1	1	1	1	1	1
	2	1	1	1	1	1	1	1	1	1	2	2	2	2	2	2	1	1	1	1	1	1
	3	1	1	1	1	1	1	1	1	1	2	2	3	2	2	3	1	1	1	1	1	2
2	1	2	2	3	2	2	3	2	2	3	3	3	3	3	3	3	2	2	2	2	3	3
	2	2	2	3	2	2	3	2	3	3	3	4	4	3	4	4	3	3	4	2	3	4
	3	3	3	4	2	2	3	3	3	3	3	4	4	4	4	4	4	4	4	2	3	4
3	1	1	1	1	1	1	1	1	1	2	3	3	3	4	4	4	1	1	1	1	1	1
	2	2	2	3	1	1	1	1	1	2	4	4	4	4	4	4	3	3	3	1	1	1
	3	2	2	3	1	1	1	2	3	3	4	4	4	4	4	4	4	4	4	1	1	1
4	1	2	3	3	2	2	3	2	2	3	4	4	4	4	4	4	4	4	4	2	3	4
	2	3	3	4	2	3	4	3	3	4	4	4	4	4	4	4	4	4	4	2	3	4
	3	4	4	4	2	3	4	3	3	4	4	4	4	4	4	4	4	4	4	2	3	4

INTERPRETATION OF THE RESULT
1 - No actions required
2 - Corrective actions required in the near future
3 - Corrective actions should be done as soon as possible
4 - Corrective actions for improvement required immediately

Fig. 1. OWAS evaluation and result interpretation.

https://doi.org/10.1051/shsconf/20184902013

2.3 Additional Data

Besides the captures of posture while working, some other primary data were also collected in the analysis, such as quantity of sweet whey powder, sack dimensions, and warehouse dimensions. Deep interview with the workers were also conducted to gather their complaints about the manual material handling process in existing condition, as well as ideas in minimizing the complaints in order to improve the quality of working.

2.4 OWAS Data Compilation

Posture data of each working movements while moving the sweet whey powder were collected as picture (.jpg) and video (.mov) using camera. There were 16 piles of powder sack in the warehouse. However, three heights of sack pile were chosen to represent the variety of activities, i.e. low, medium, and high. Each capture of body parts was then graded using the OWAS action, as listed in Table 1. Video data were used for helping the grading process. If there were more than one position of body part, the higher number of action level was chosen. OWAS scores were calculated and categorized using ErgoFellow 3.0 software, which was developed by FBF Sistemas Brazil.

2.5 Tool Design Apparatus

After examining the OWAS scores of existing conditions, a working tool was then designed for helping the handling process, in order to minimize the OWAS scores. Design process of the working tool was carried out using CATIA V5, a computer-aided design software by Dassault Systèmes French.

3 Results and Discussion

3.1 Existing Condition

Some pictures of the handling process were shown in Fig. 2. As stated before, the activities captured are for stacking the low pile, medium pile, and high pile of powder sack.

Fig. 2. Postures of handling process.

The OWAS scores for each pictures were then identified. When more than one positions were captured, the highest score was selected. The results are summarized in Table 2.

Table 2. OWAS scores of handling process.

Pile	Activity	Digit	Interpretation
Low	Lifting	3	Corrective actions should be done as soon as possible
	Carrying	2	Corrective actions required in the near future
	Lowering	3	Corrective actions should be done as soon as possible
Medium	Lifting	3	Corrective actions should be done as soon as possible
	Carrying	2	Corrective actions required in the near future
	Lowering	3	Corrective actions should be done as soon as possible
High	Lifting	3	Corrective actions should be done as soon as possible
	Carrying	2	Corrective actions required in the near future
	Lowering	3	Corrective actions should be done as soon as possible

As displayed above, all activities of moving the sweet whey powder shown high risks of musculoskeletal disorders. As a summarize, the handling process needs corrective actions as soon as possible. The results were based on the poor condition of lifting, carrying, and lowering the load, whether in low pile, medium pile, or high pile of powder.

In the lifting process, the workers bent their back while reaching the load. Both of their arms were in the lower position than the shoulders. They were also standing on bended knees. Additionally, the load they lifted was more than 20 kilograms.

In the carrying process, the workers back were straight, while both arms were higher than the shoulders position. Workers were walking through the process, and carrying more than 20 kilograms of load.

In the lowering process, the workers bent their back when lowering load in the low and medium height of sack pile, while in the high pile they can kept the back straight. On the low and medium height of pile, both arms were lower than the shoulders. Both arms were higher than the shoulders when they stacked in high pile. On all height of pile, they were standing on one bent leg, with a more than 20 kilograms of load.

As known widely, bending posture was highly correlated with back injury. Someone's back muscles are in stretching position when he is bending his back. The discs of spines encounter high pressure, including the nerves inside. When this bending position happening repetitively, the discs will be weakened, and/or ending in disc rupture, or called herniation [7]. Symptoms of this condition were include back pain, strain, and cramp.

In addition, workers' arms were higher than their shoulders in some activities. In carrying process, moreover, their heads also supported the load along the process. This condition will possibly end up in neck stiffness, neck cramp, and also neck injury.

3.2 Tool Design

After considering the workers postures along the manual material handling and also their ideas of improving the process, a trolley was chosen. Fig. 3 displays the design of this trolley.

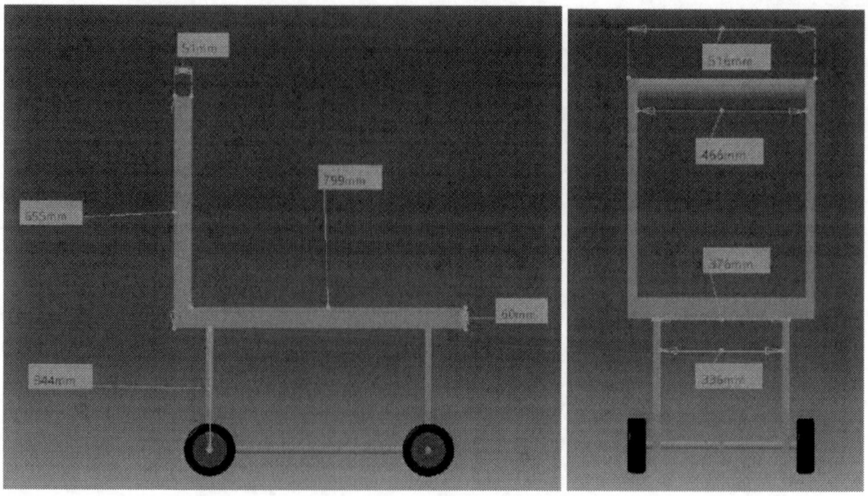

Fig. 3. Trolley design (side and front view).

The trolley was designed using Indonesian anthropometry dimensions, applying 95 percentile of men. The trolley height was 1074 mm, equivalent with Indonesia populations elbow height. The diameter of the handle was also adjusted to Indonesian hand grip, which was 51 mm. The height of sack base was 544 mm, equivalent with Indonesian anthropometry knee height dimension.

3.3 Improved Condition

Applying the trolley proposed, analysis of working postures using OWAS was made once again. Fig. 4 shows the lifting and lowering process with CATIA software, while Fig. 5 presents the carrying process.

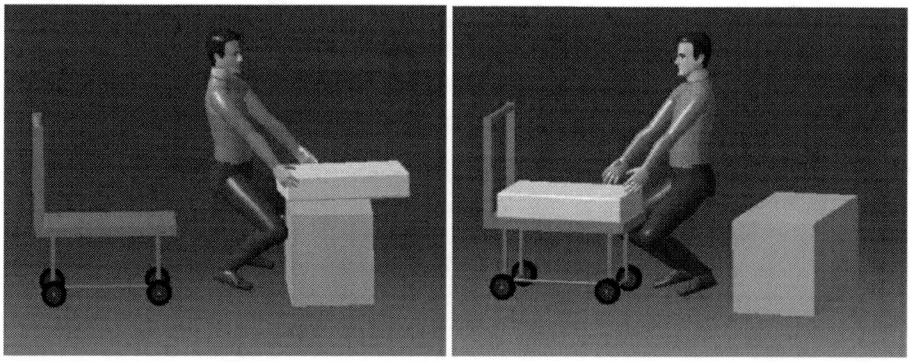

Fig. 4. Lifting and lowering load with trolley.

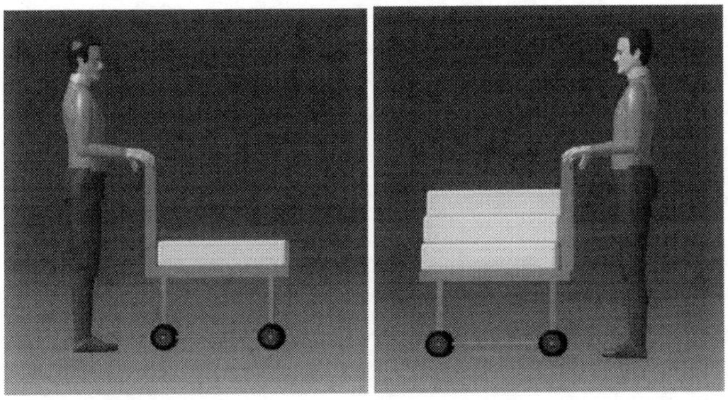

Fig. 5. Carrying load with trolley.

The lifting and lowering process with trolley resulted in OWAS score of category 2 (Corrective actions required in the near future), lower than the existing condition. Surprisingly, the carrying process was resulted in OWAS score of category 1 (No actions required), much lower than the existing condition.

Furthermore, the handling activity using trolley will be more efficient than the existing condition, since workers can carry two to three sacks at one time. The more efficient handling process means the faster workers finish their tasks, the more kind of tasks they can handle, the more saving company can make, and hopefully the more profit company can get.

Additionally, to maintain the minimum musculoskeletal risks during material handling, a base of 544 mm of height need to be provided. The maximum number of stacking suggested is ten. Therefore, the OWAS score could be maintained in category 1 or 2, which is consider low risks.

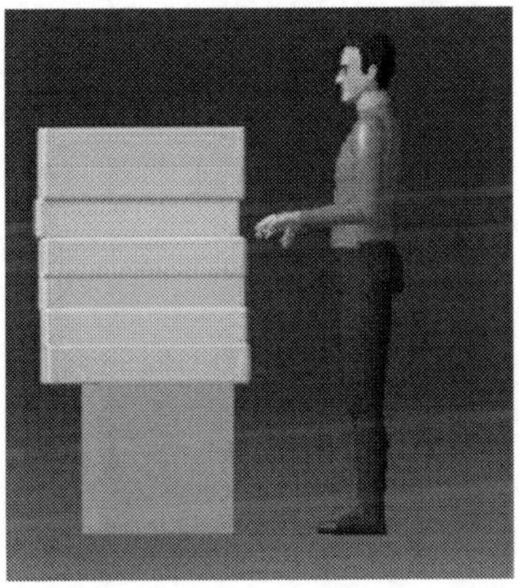

Fig. 6. Pile of powder sacks with proposed base.

4 Conclusion and Further Study

The existing condition of unloading process of sweet whey powder in CV Cita Nasional were carried out in high musculoskeletal risks, assessed by the worker postures using OVAKO Working Posture Analysis System (OWAS). Therefore, a trolley was designed in Indonesian anthropometry dimensions to help minimize the risks. Later, the lifting, carrying, and lowering activities with the proposed tool could be conducted in lower OWAS score, interpreted as lower musculoskeletal risks.

Further study can be performed on other type of work that involving manual handling using the procedure in this study. More anthropometry data could be involved in the future study, applying Indonesian population from [8]. Additional biomechanics analysis can also be conducted on the present object, for example using electromyography (EMG) method as in [9], to make sure that the tool designed is significantly minimizing the MSDs risks.

References

1. A. M. Basahela, Investigation of work-related Musculoskeletal Disorders (MSDs) in warehouse workers in Saudi Arabia. *Procedia Manufacturing*, **3**, 4643-4649 (2015)
2. Canadian Centers for Occupational Health & Safety, *WMSDs*, http://www.ccohs.ca/oshanswers/diseases/rmirsi.html, Retrieved on 7 October 2016 (2014)
3. E. J. Wright, R. A. Haslam, "Manual Handling Risks and Controls in A Soft Drinks Distribution Centre", *Applied Ergonomics*, **30 (4)**, 311-318, (1999)
4. NIOSH, *Musculoskeletal Disorders and Workplace Factors: A Critical Review of Epidemiologic Evidence for Work Related Musculoskeletal Disorder*, NIOSH: Centers of Disease Control and Prevention (1997)
5. NIOSH, *Ergonomic Guidelines for Manual Material Handling*, NIOSH Publications Disseminations, Columbia (2007)
6. O. Karhu, "Correcting working postures in industry: A practical method for analysis", *Applied Ergonomics*, **8(4)**, 199-201 (1977)
7. D. B. Chaffin, G. B. J. Andersson, B. J. Martin, *Occupational Biomechanics*, John Wiley & Sons, New York (1999)
8. A. Widyanti, M. Mahachandra, A. Wranitisandhini, A. Rahim, "Development of web based information system for Indonesian anthropometry data", *Proceeding of APCHI-ErgoFuture 2010*, 669-673 (2010)
9. B. Widjasena, P. Sinaga, H. Prastawa, "Participatory approach in designing tea picker bag to reduce musculoskeletal disorders at Medini Farm, Limbangan District, Kendal Regency", *Proceeding of ErgoFuture 2006: International Symposium on Past, Present, and Future Ergonomcis, Occupational Safety and Health*, 65-69 (2006)

Individual based simulation for online marketplace diffusion among trading small medium enterprises: A conceptual framework

Singgih Saptadi[1,2,*], *Arlita Rahma* Widyasrini[1] , *Bonita Melinda* Pangaribuan[1] , Sriyanto[1] and *Ary* Arvianto[1,2]

[1]Industrial Engineering Department, Diponegoro University, Semarang, Central Java, Indonesia
[2]Logistics and E-Business Research Center, Diponegoro University, Semarang, Central Java, Indonesia

Abstract. The number of Indonesian SMEs and Indonesian Internet users are increasing. This situation is of concern to the Indonesian government. The government then launched a SME go online program with a target of 8 million SMEs using the online platform in 2019. One of the online platforms that SMEs may use in Indonesia is the online marketplace such as Tokopedia, Bukalapak and others. Previous research studied the factors that support or hinder the adoption of e-commerce in SMEs environment but had not explored how the interaction of those factors in the adoption of online marketplace. In a social system, the diffusion and adoption of online marketplaces in SMEs is influenced by socio-economic characteristics, personality variables and communication behaviours that exist and are acquired by individuals. This paper describes a conceptual framework for the development of an individual-based simulation commonly referred to as an agent-based simulation to understand the adoption by SMEs and the diffusion of online marketplace in SME environments.

1 Introduction

Small and medium enterprise (SME) is a labor-intensive company with requirements such as education, worker skills, capital or technology tend to be simple [1]. Indonesian Ministry of Cooperatives and Small Medium Enteprises stated that the number of SMEs in Indonesia in 2013 amounted to 57,800,000 units (99.9% of business enterprises in Indonesia). With that amount, SMEs have huge contributions to Indonesian economy within last five years that are 60.34% of Product Domestic Brutto and 97.22% of national employment (http://www.kemenperin.go.id/artikel/14200/Kontribusi-UMKM-Naik).

Internet has a very rapid user growth. Currently, internet penetration in Indonesia reached 53.7% of population [2] and this means more than 140 million internet users in Indonesia. With a huge and growing market potential, internet use by SMEs is a concern of Indonesian government. Considering the contribution of SMEs to Indonesian economy and the economic potential of the internet, the government launched a SMEs go online program

* Corresponding author: singgihs@ft.undip.ac.id

with a target of 8 million SMEs utilizing online platforms in 2019 [3]. An online platform that is feasible to be utilized by SMEs freely and easily is marketplce online in Indonesia. Some of the known online marketplace are Tokopedia, Bukalapak and others. Unfortunately, until 2017 only 3.79 million SMEs that utilize the online platform.

The online marketplace is a form of e-commerce that is often referred to as multichannel ecommerce. The development of internet technology enables the emergence of new varieties of e-commerce. As an innovation, the online marketplace is perceived as something new [4] in marketing SME products. The diffusion of online marketplace in the SMEs environment is influenced by various factors, both the supporting factors and the inhibiting factor of online marketplace adoption by SMEs..

Some researches studied the factors that support or hinder the adoption of e-commerce, especially in the SMEs. MacGregor & Vrazalic studied factors that drive e-commerce adoption by SMEs in Wollongong (Australia) and Karlstad (Sweden) [5]. These factors are: consumer demand, increased sales, and increased control of goods. Govindaraju & Chandra studied the inhibiting factor of e-commerce adoption by SMEs in Indonesia and showed that human resource and sources of information as two inhibiting factors [6]. Saptadi et al examined the influence of SME owners' support and SMEs IT sophistication on IT adoption by SMEs [7].

These studies have not explored how the interaction of various factors in the adoption of online marketplace. The diffusion of the online marketplace, as well as the diffusion of other innovations, occurs in a social system. In social systems, the diffusion of the online marketplace in SME environments and the adoption of online marketplace by an SME is influenced by social factors and individual characteristics of the adopters. Based on this, the study sees the need for an understanding of the adoption of online marketplaces by SMEs and the diffusion of online marketplaces among SMEs through a model that illustrates the individual characteristics of the adopter and of the social system. We therefore examine the development of an individual-based model commonly referred to as an agent-based model to understand the adoption by SMEs and the diffusion of online markeplace in SME environments.

This paper explains our initial idea of individual-based modeling for online marketplace diffusion. In the next section, we describe some concepts related to the diffusion of innovation in the social systems that mostly we adapt from Rogers [4]. Then, we explain the conceptual framework that we use as the basis for the next activity in this study. This paper concludes with an explanation of our next research agenda.

2 Difusion of Innovation and Individual Based Modeling

Diffusion is a process of an idea or innovation communicated or dispersed through certain channels at a time period among members of its social system. Innovation is an idea or practice that is perceived new by individuals who will adopt it or apply it [4]. Messages or information disseminated in the process of interaction or communication containing ideas, ideas or practices that are considered new. According to Rogers, the diffusion of innovation is influenced by four elements: the innovation itself, communication channels, time, and social systems [4].

Researchers understood that individual decision about an innovation is not an intantaneous act. The adoption decision process of an innovation by a decision maker unit is an individual process where the adopter knew the innovation, got additional knowledges about the innovation and followed by accepting or rejecting the innovation [4]. Individual decision-making process for adopting an innovation are influenced by socio-economic characteristics, personality variables and communication behaviours of individuals. An innovation, like marketplace online, enters a social system which has a unique socio-economic structure. an online marketplace is also exposed to individuals who have diverse personalities and communicating behaviors [4]. Considering the online marketplace will be adopted by SMEs that have various characters and diffused in the social environment of SMEs, then understanding the diffusion of the online marketplace in SMEs requires methods that consider the characteristics of individuals that live within a social system.

Individual-based simulation is a form of computational modeling in which a phenomenon of the agent and its interactions is modeled. The modeled phenomena include the characteristics of agents, environments, and descriptions of interactions between agents and agents with their environment [8]. The answer is that in ecology, the individuals are not atoms but living organisms. Individual organisms have properties an atom does not have. Individuals grow and develop, changing in many ways over their life cycle. Individuals reproduce and die, typically persisting for much less time than the systems they belong to. Individuals need resources, so they modify their environment. Individuals differ from each other, even within the same species and age, so each interacts with its environment in unique ways [9].

3 The Conceptual Framework

In this section, we discuss the conceptual framework underlying this research. The first part of the conceptual framework is the definition of an agent. The agent or entity in this study is the decision maker in SMEs (we shortly name as SMEs), and the government. An SME interacts with other SMEs through conversations with other SMEs, news from mass media or facilitated training by the government.

As an agent, each SME has individual attributes, namely (1) adoption status as adopter or non-adopter, (2) the initial value of any inhibiting factors and drivers. In this study, we involve some of the following factors derived from previous studies: (a) Have educated human resources; (b) Knowledge of the benefits of online marketplaces; (c) Knowledge of implementation of online marketplaces; (d) Experience using electronic marketing media; (e) Have time to learn an online marketplace; (f) Have time to operate an online marketplace; (g) Online marketplace compliance with business processes; (h) Have support from management to use online marketplace; (i) the threat from competitors who are using online marketplace; (j) Get information about online marketplace from media; (k) Consumer order through online marketplace; (l) Sales increase; (m) Improved control and follow-up.

The value of these variables will change with the interaction of SMEs with other agents as well as news from the mass media. Based on the values of these attributes, SMEs may accept or reject the adoption of an online marketplace.

Process overview is the second part of this conceptual model. Process overview discusses the sequence of activities that agent will perform and when the individual attribute will be updated [10]. Fig. 1 illustrates the process overview of this model. First channel in diffusing marketplace online is agent-to-agent interaction. Initial simulation will initialize the state variable of each agent. The adopter agent as will be a partner on the interaction of the conversation. The non-adopter agents as potential adopter, if interacting with the adopter will get additional value each individual attribute (drivers or inhibiting fators of adoption). If the value of each of the driver and inhibiting factors of online marketplace adoption is greater than the average adopter value of each factor, then agents with potential status will be satisfied with their interactions and then they may become new adopter or stay as non-adopter.

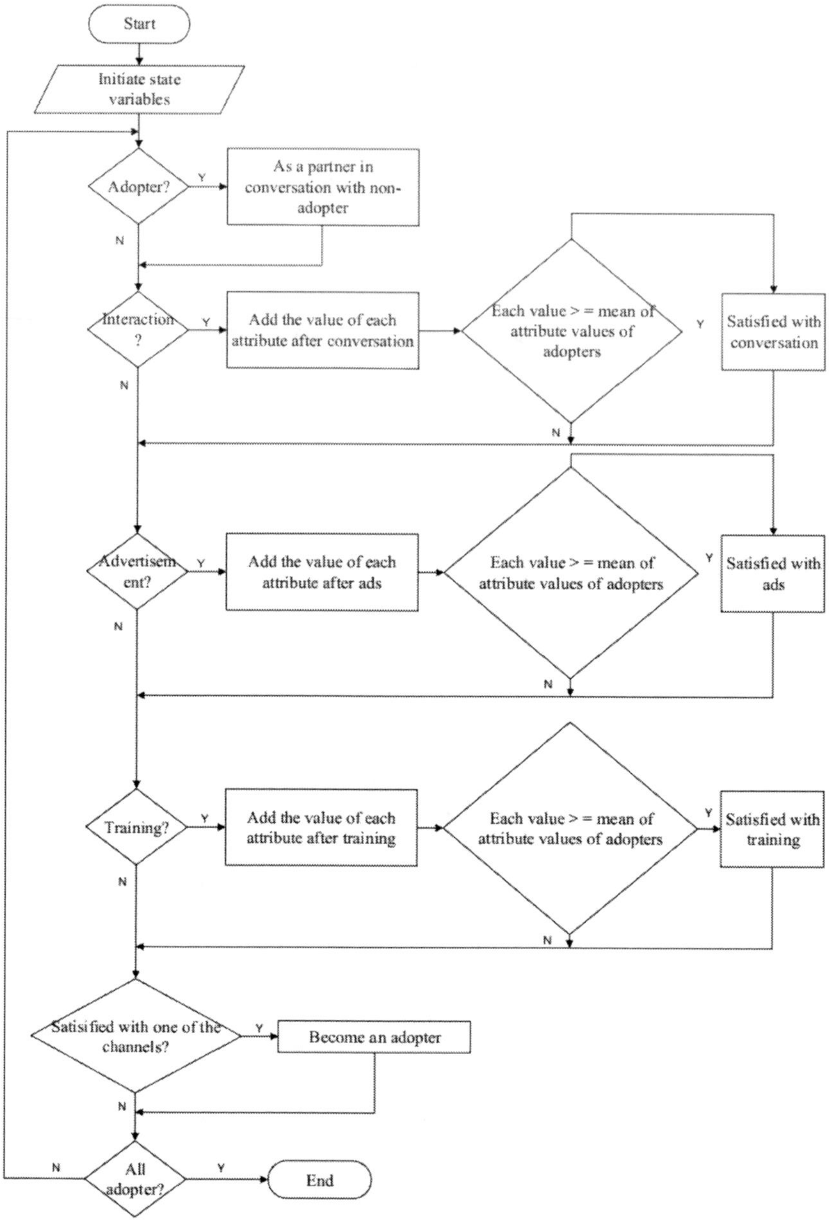

Fig. 1. Process overview.

Second channel in diffusing online marketplace is news from mass media. If an agent with a potential status receives information about an online marketplace from mass media or advertising, it will receive an additional value of the online marketplace and if this value is greater than the marketplace's online information adopter value of the ad, then the agent is satisfied with the advertisement and they may become new adopter or stay as non-adopter.

The third channel is facilitated training by government. If an agent with a potential status receives information about the online marketplace of the training, it will receive a value-added online marketplace information from the training and if the value is greater than the value of the online marketplace information adopter of the training, then the agent is satisfied with the training and they may become new adopter or stay as non-adopter.

4 Next Agenda

This paper presents a conceptual framework for the development of individual-based simulations (agent-based models) from the diffusion of the online marketplace. The conceptual framework includes defining agents / entities and developing process overviews. This paper describes UKM as an agent that has several attributes. The agent attributes include the current status of adoption and the factors that influence the adoption of an online marketplace by SMEs. The process overview explains how SME attributes shift the status of SMEs from non-adopters to adopters.

The conceptual framework will be used in this research. A questionnaire will be developed and distributed to trading SMEs. The results of data collection will be processed and elaborated to obtain any patterns of characteristics of SMEs in adopting the online marketplace.

The definitions of agents and their attributes and characteristic patterns of SMEs in adopting will be translated into agent-based simulation languages. To date, we have not determined the software for agent-based simulation programming. based on popularity, Netlogo is the strongest candidate to be chosen. Our simulation results will be analysed to gain a deeper understanding of the diffusion of online marketplace in the SME trading environment.

References

1. D. D. Utama, Darwanto, *Pengembangan Usaha Mikro Kecil Dan Menengah (UMKM) Berbasis Ekonomi Kreatif Di Kota Semarang*. http://eprints.undip.ac.id/40407/ (Diponegoro University Institutional Repository, 2013)
2. Internet World Stats, *Internet Usage in Asia*. https://www.internetworldstats.com/stats3.htm. (2017)
3. K. Ayuwuragil,. *Kemenkop UKM: 3,79 Juta UMKM Sudah Go Online*.https://www.cnnindonesia.com/ekonomi/20171115161037-78-255819/kemenkop-ukm-379-juta-umkm-sudah-go-online. (CNN Indonesia, 2017)
4. E. M. Rogers, *Diffusion of Innovations (5thed.)*. (The Free Press, NewYork, 2003)
5. R. MacGregor, L. Vrazalic, *Electronic Commerce Adoption in Small to Medium Enterprises (SMEs) A Comparative Study of SMEs in Wollongong (Australia) and Karlstad (Sweden)*. https://www.uow.edu.au/content/groups/public/@web/@commerce/@econ/documents/doc/uow012275.pdf. (2004)
6. R. Govindaraju, D. Chandra, IEEE 3rd ICCSN, *E-commerce adoption by Indonesian small, medium, and micro enterprises (SMMEs): Analysis of goals and barriers*, 113-117 (2011)
7. S. Saptadi, I. Sudirman, T.M.A A. Samadhi, R. Govindaraju, *Owner's Support, IT Sophistication and IT Adoption in Indonesian Manufacturing SMEs*, (2015)
8. U. Wilensky, W. Rand, *An Introduction to Agent-Based Modeling,* (The MIT Press, London, 1955)
9. S.F. Railsback, V. Grimm, *Individual-Based Modeling and Ecology,* Princeton University Press, Princeton, 2005)
10. V. Grimm, U. Berger, D.L. DeAngelis, J.G. Polhill, J. Giskee, S.F. Railsback, EM, *The ODD protocol: A review and first update*, **221**, 2760-2768 (2010)

Cost-benefit analysis of flight extended operations (ETOPS) for Garuda Indonesia airways

Ratna Purwaningsih[1], *Lakshita* Pritandari[1,*], *Haryo* Santoso[1]
[1]Industrial Engineering Department, Diponegoro University, Jl Prof Soedharto, SH, Tembalang, Semarang, 50275 Central Java, Indonesia

Abstract - Garuda Indonesia reducing the flight cost used flight extended operations (ETOPS). ETOPS can reduce travel time and fuel consumption. The cost-benefit analysis was conducted to compare the flight between NON ETOPS flight and ETOPS flight Cengkareng - Perth - Cengkareng route. Net benefit of ETOPS flight is USD 1.212.863 and NON ETOPS is USD 1.154.894. Cost structure analysis was conducted to identify the percentage of flight cost component. The biggest percentage of cost was direct flight cost. It is equal to 49,53% for route Cengkareng - Perth NON-ETOPS and 47,70% for ETOPS. While for the route Perth - Cengkareng NON-ETOPS and ETOPS have the same amount of 46.03%. Based on the results of the cost-benefit analysis, it is evident that the ETOPS flight can reduce the fuel cost, although the flight requires trained pilots. Contribution of the paper is brief describe on the structure of revenue and expenditure items in airways business. The structure is specific, different from other transportation business.
Keywords: ETOPS, fuel, cost-benefit analysis, cost structure

1. Introduction

Determination of public transport fares require the right policy to bridge the interests of passengers as consumers and employers as providers of public transport services. Many factors affect the transport tariffs, such as public purchasing power, the cost of maintenance, spare parts price, fuel prices, facilities and infrastructure [1]. Air transportation has specific structure on revenue and expenditure items which not much discuss on scientific publication. Technical knowledge is needed to obtain the data of revenue and expenditures of airways business.

Garuda Indonesia is the largest airline in Indonesia. Garuda Indonesia airlines carry more than 25 million passengers annually, serve flights to more than 40 domestic destinations and 26 international destinations. Garuda Indonesia has prepared various aircraft to meet the needs of consumers for international flights, such as Airbus Industries A330-300, Boeing 737-800, and others. Boeing 737-800 is one type of aircraft with seat load factor of 162 seats, this type of aircraft can be used to travel internationally to countries such as Perth, Australia.

[*] Corresponding author: prita1803@gmail.com

ETOPS or Extended Operations or formerly known as Extended Range-Twin Engine Operations is a technology carried out by a two-engine aircraft which has the ability to fly only with one machine for a certain period on the scope of an adequate airport (adequate airport). ETOPS can reduce the gasoline consumption due to shorter routes and leads to cost savings. Now, Garuda Indonesia is conducting a feasibility analysis of ETOPS technology for Boeing 737-800 on the Cengkareng - Perth flight route using New Generation aircraft fleet.

ETOPS flights require trained pilots, it is the investment cost. Management needs to know whether ETOPS flight is still feasible or not by calculating all cost involved. The calculation of feasibility analysis of ETOPS and NON ETOPS be reviewed by cost-benefit analysis. Cost-benefit analysis can be used to make decisions from several existing investment alternatives by reviewing the costs associated with the object under investigation. This study does a review NON ETOPS and ETOPS flights for Cengkareng - Perth - Cengkareng route based on flight cost plan. The study aims to compare the resources uses for flight on structure analysis and calculating the cost-benefit analysis. The ETOPS flight used in this study was ETOPS 90 minutes.

2. Literature review

ETOPS or Extended Operations is an operation performed by twin-engine aircraft that have the ability to fly for 60 minutes or more with one machine in dead from the nearest airport in certain routes [2]. An adequate airport is a decent airport to serve as the launching point of the ETOPS aircraft [3]. The operating area of ETOPS is the area that the relevant aircraft can fly under the ETOPS regulation. This area is characterized by a circle centered on the adequate airports. ETOPS flights alone can save the fuel consumption and reduce the flight cost. In the USA, more than 5.5 million ETOPS twinjet flights have been logged worldwide since 1985, and every day some 143 operators perform 1,750 more these operations set the highest standard for safe, reliable long-range flying or RLRF [4].

The cost structure is good to describe all the costs required to run a business model [5]. Identification the percentage of flight cost will help to describe the impact of saving fuel consumption on ETOPS and NONETOPS flight. Some cost component occurred while preparing the ETOPS flight. Costs are accounted for major activities, key resources, and partnerships determined. Aviation cost of differs slightly from the operational cost of manufacture companies. The cost consists of direct cost and indirect cost. Direct cost or operational costs is a cost associated with the aircraft and flight cost. The cost incurred by airlines that dependent in the intended destination, aircraft technology used, flight duration and other related aspects. Indirect cost is the cost for daily aircraft operations, aircraft and flight administration, and overhead cost (costs incurred excluding operational costs).

All these costs will be processed to get the value of the total cost. Total cost is the accumulation of all the costs required for the associated route. Total cost consists of the incremental cost of flight interrupted and HO administration. Flight interrupted is the cost incurred when an unavoidable delay occurs. While the administration of HO or administration head office is the cost of administration in the main office to handle the administration of the related routes. Morlok (1988) stated that the management of transport businesses face a very wide choice in terms of pricing and operating plans [6].

Cost-benefit analysis or CBA is a practical tool for assessing the feasibility of a project. It is important to consider the project over the long term such as reciprocity to consider the impact of a project on stakeholders [7]. CBAs on business units or those used within a company consider the commercial aspect of a project and regardless of the social effect [8]. The advantages of CBA is more efficient on using of economic resources, and the use of funds monitored appropriately. The disadvantages of CBA is that the impact of the project cannot be analyzed precisely and many factors influence lead to increase cost [9].

3. Research Methods

There are five stages to perform cost-benefit analysis calculation [10]. First, define a problem and identify possible options/alternatives. Identify problems that occur by getting information from relevant parties and start looking for solutions to solve the problems at hand. The solution offered consists of several alternatives, the best alternative will be selected based on cost-benefit analysis calculation. Second steps are determined inputs and outputs (impacts) of each option. All aspects that affect the calculation of cost-benefit analysis should be detail reviewed accordance to each available alternative.

Third steps is the valuation of the benefits and costs of each option ('net benefits'). The calculation of income and expenditure is calculated on the basis of alternatives previously reviewed, after which the profit of each alternative is calculated. Step four is to identify the 'best' option based on the calculation of each of the advantages that have been done. The last step is selected the best alternative that provides problem-solving and the benefits to be desired.

4. Result and Discussion

The first stage of cost-benefit analysis is defined problem and identify possible options or alternatives. Based on experienced, the demand for Cengkareng - Perth - Cengkareng route is quite attractive to consumers, indicated by high seat load factor of 75.96% for the Cengkareng - Perth route and 77.80% for the Perth - Cengkareng route. But economic conditions in 2016 are in a critical position, where the revenue of this airline has decreased compared to previous years. All program for savings already made by Airlines to increase the profit. Extended Operations or ETOPS flights on Cengkareng - Perth - Cengkareng route using Boeing 737-800 New Generation aircraft can be one of the saving programs. This stage described data of flight plan for Non-Etops and Etops based on the flight plan described in Table 1.

4.1. Cost-benefit analysis

The second stage of cost-benefit analysis is determining inputs and outputs of each option, means identify the revenue and expenditure structure of airways business. Based on financial statements, the revenue earned from NON-ETOPS Cengkareng - Perth flight in 2016 was USD 5,910,056 and Perth - Cengkareng was USD 6,836,053. The financial statements are recorded for each flight made by Garuda Indonesia, which consists of net-income (revenue and expenditures). The nett revenue was obtained with load factor 66.02% for Cengkareng - Perth flight and 66.40% for Perth - Cengkareng flight. Load factor is total of passengers, cargo, mail, and baggage loaded divided by the available capacity. The revenue resulted from 191 times total flights for Cengkareng - Perth and 191 for Perth – Cengkareng. The calculation for ETOPS flight revenue for the same route made use the same load factor and flight frequency. A detail description of the revenue for the Non-ETOPS flight is listed in Table 2.

Table 2 explain the structure of revenue on airways business. The revenue of ETOPS and NON ETOPS flight give the same result. The factor give highest influence on revenue estimation is the load factor assumption. That's why the revenue of ETOPS and NON

ETOPS should not compared. The data of revenue only needed to calculate the benefit as difference between revenue and cost.

Table 1. Comparison of Etops and Non-etops Flight information 2016

	CENGKARENG - PERTH				PERTH – CENGKARENG				
	Non-etops	Etops	Gap (flight)	Gap (year)	Non-etops	Etops	Gap (flight)	Gap (year)	unit
Payload	15.144	15.578	434	83.328	17.179	17.510	331	63.552	kg
Duration	269	262	7	1.344	279	275	4	768	minute
Trip fuel	10.701	10.326	375	72.000	11.522	11.267	255	48.960	kg
	13.590	13.114	476	91.440	14.633	14.309	324	62.179	liter
Block fuel	16.920	16.488	432	82.944	15.706	15.376	330	63.360	kg
	21.488	20.939	548	105.34	19.947	19.528	419	80.467	liter
Distance	1.825	1.730	95	18.240	1.830	1.800	30	5.760	Nm
	3.379	3.203	176	33.780	3.380	3.204	176	33.780	Km

Table 2. Revenue of Non-ETOPS Flight 2016

Revenue	CGK – PERTH (US $)	PERTH – CGK (US $)
Passenger Revenue Gross	5.625.247	6.628.417
Excess baggage revenue	18.191	18.501
Passenger Revenue discount	3.460	69.195
Passenger Revenue net	5.639.979	6.577.723
Freight revenue gross	140.953	7.162
Freight revenue discount	97.382	4.596
Freight revenue net	43.571	2.566
Mail revenue	1.673	0
Other revenue	224.833	255.765
Net Revenue	5.910.056	6.836.053

The flight cost or expenditure consist of many kinds of cost describe on Table 3. Direct traffic cost consist of passenger commission, freight commission, credit card commission, catering, on board service, and reservation cost paid by customers. Direct flight cost is the highest one.

The third stage is the valuation of the benefits and costs of ETOPS flight and Non-ETOPS flight. Cost planning required for Non-Etops and Etops for Cengkareng - Perth - Cengkareng route was planned and revenue obtained in 2017 estimated. Several assumptions are 192 times flights in a year and 80% load factor. Calculation of Etops and Non-ETOPS flight cost for Cengkareng - Perth - Cengkareng route only different on direct flight cost, fleet cost and pilot training cost. Direct flight costs on Non-Etops for Cengkareng - Perth route is USD 2.429.000 and ETOPS flight cost USD 2.391.000

resulting a gap of USD 37.600. Direct flight cost for Perth - Cengkareng route Non-Etops was USD 2.834.000 and the Etops cost was USD 2.813.000 resulting gap of USD 20.000. Fleet cost only changed at the depreciation cost and lease cost of aircraft, while for the cost of insurance does not change. Fleet cost flight Non-ETOPS route Cengkareng - Perth was USD 1.266.000 and ETOPS flights cost USD 1.246.000 thus spending a gap of USD 20.000. Perth - Cengkareng flight route the Non-ETOPS fleet cost was USD 1.295.000 and Etops was USD 1.283.000 resulting in a gap of USD 11.600.

Table 3. Comparison of Etops and Non-Etops Flight cost 2016

Cost	CGK – Perth		Pert - CGK	
	NON ETOPS (USD)	ETOPS (USD)	NON ETOPS (USD)	ETOPS (USD)
Direct Traffic Cost	774.963	774.963	954.956	954.956
Direct Flight Cost	2.428.913	2.391.251	2.833.703	2.833.703
Fuel Aircraft	1.203.150	1.140.5	1.203.6	1.203.648
Cockpit Crew	152.466	149.978	153.772	153.772
Cabin Crew	182.990	180.004	178.236	178.236
Landing	174.496	174.496	126.835	126.835
Handling	196.978	196.978	610.744	610.744
Air Traffic	86.570	73.786	105.300	105.300
variable	432.259	475.485	455.164	455.164
Indirect Cost	205.855	205.855	213.809	213.809
Fleet Cost	1.266.213	1.245.8	1.295.3	1.283.720
Depreciation	93.559	92.033	95.653	94.781
Lease Aircraft	1.152.300	1.133.4	1.179.2	1.168.539
Insurance	20.354	20.354	20.399	20.399
Overhead Cost	396.218	396.218	437.631	437.631
Flight Interrupted	4.376	4.376	4.549	4.549
Administration HO	358.697	358.697	415.996	415.996
Training Cost	-	5.819	0	5.819
TOTAL COST	5.435.235	5.383.0	6.155.9	6.150.183

Training cost is used for 1 pilot of Boeing 737 - 800 New Generation. The Cengkareng - Perth route requires 25% of the total training cost. The pilot training cost for ETOPS flight is IDR 5 million for each person. Cengkareng - Perth - Cengkareng flight route alone requires 5% of the total 310 crew of Boeing 737 - 800 New Generation or 31 crew. Total training cost for Cengkareng flight - Perth - Cengkareng for 31 crew is IDR 155 million or if converted into dollars is USD 11,638.

The fourth stage compares the net benefits of each option. Cengkareng - Perth route revenue is USD 7.217.054 per year, then net benefit Non-ETOPS flight is USD 1.782.000 and ETOPS flights is USD 1.834.000. Perth - Cengkareng route gain potential revenue USD 8.291.000 per year and obtained a net benefit for Non-ETOPS flight USD 2.135.000 and Etops USD 2.141.000. The fifth stage is to identify the 'best' option. The choice of best option can be seen based on the largest net benefit from each route. Based on the largest net benefit for the Cengkareng - Perth - Cengkareng route, Etops is chosen as the best option compared to Non-ETOPS flights. Fuel that can be saved if using ETOPS is 185,806 liters per year or 192 times flight. Total cost saved is USD 57.969 per year. Table 5 outlines the overall flight costs of Non-Etops and ETOPS routes Cengkareng - Perth - Cengkareng.

Based on financial statements, the revenue earned from NON-ETOPS Cengkareng - Perth flight in 2016 was USD 5,910,056 and Perth - Cengkareng was USD 6,836,053. Total revenue was USD 12.746.109. Table 5 give the total flight cost for NON ETOPS and for ETOPS. From the data we can make a calculation of benefit of flight and compare the benefit between ETOPS and NON ETOPS. Cost-benefit analysis shows that ETOPS flights are more profitable, although requires training costs for pilots. The benefit for 2016 NON ETOPS flight was 1.154.894 USD and for ETOPS was 1.212.863 USD. Although the used of ETOPS flight on 2016 not increase the benefit significantly, the flight for the route was increase in 2017. The estimation of load factor was 80 % compare to 2016 60 %. In this case, the effect of use ETOPS flight will give profit increase significantly.

4.2. Cost Structure Analysis

The most influential cost of transportation business is fuel cost, it also happens on bus operational costs [1]. The result of cost structure analysis for Cengkareng-Perth route shows that direct flight cost is the biggest expenditure components for Non-ETOPS flight (44.69%) and fleet cost are the biggest expenditure components for ETOPS flight (44.42%). Etops flight decreased the direct flight cost 0.27%, fleet cost 0.16%, and need additional training cost 0.11% of all cost. The result for Perth - Cengkareng route also shows that direct flight cost and fleet cost as the largest cost components for Non-Etops and Etops. Etops flight result on a decrease of direct flight cost 0%, fleet cost 0.17%, and need additional training cost by 0.09%. The most influential resource on Non-Etops and ETOPS flights is the fuel cost, describe Table 4.

Table 4. Comparison of Etops and Non-Etops Cost Structure

	CGK - PER		PER – CGK	
	Non etops	Etops	Non etops	Etops
Direct Traffic Cost	14,26%	14,40%	15,51%	15,53%
Direct Flight Cost	44,69%	44,42%	46,03%	46,03%
Indirect Cost	3,79%	3,82%	3,47%	3,48%
Fleet Cost	23,30%	23,14%	21,04%	20,87%
Overhead Cost	7,29%	7,36%	7,11%	7,12%
Flight Interrupted	0,08%	0,08%	0,07%	0,07%
Administration HO	6,60%	6,66%	6,76%	6,76%
Training Cost	0%	0,11%	0%	0,09%
TOTAL COST	100%	100%	100%	100%

Cockpit crew travel and cabin crew travel are also one of the most cost-affecting aspects of cost. These costs are part of the direct flight cost group. Non-Etops and ETOPS flights also associated with fleet costs, and training costs. The efficiency or savings made by ETOPS flight can be seen from the percentage of direct flight cost, fleet cost, and overhead cost.

Based on the details of cost structure analysis, it is proven that fuel is the most influential component of direct flight cost. Fuel consumption difference between Non-Etops and

ETOPS flight for Cengkareng - Perth route is 105 liters per year, and for Perth - Cengkareng route is 80 liters per year. Thus, it can be concluded that by using Etops result on cost reduction from fuel savings caused by flight duration. Cost saving can be used for route-related marketing costs in order to increase load factor, improved quality of passenger care and increased service for passengers.

5. Conclusions

Cost-benefit analysis shows that ETOPS flights are more profitable than NON ETOPS. Compare to NON ETOPS, ETOPS flight requires training costs for pilots but saving the fuel consumption cost. Based on cost structure analysis, the greatest expenditure is direct flight cost component due to fuel cost. Fuel consumption difference between Non-Etops and ETOPS flight according to amount of distance reduction or time flight.

Analysis on ETOPS and NON ETOPS flight can be done more sensitive if we concern only on cost reduction result of fuel consumption decreasing. The revenue calculation significantly influence by load factor assumption and the calculation of revenue will give the same value for ETOPS and NON ETOPS flight model.

References

1. R. Purwaningsih., H. Nugroho., "Analysis of Public Transport Rates Based on Vehicle Operational Costs and Consumer Willingness to Pay (WTP)", Conference proceeding ICOMS 2016, Yogyakarta, Indonesia, (2016).
2. Transport Canada Safety and Security. *Safety Criteria for Approval of Extended Range Twin-Operations (ETOPS).* Canada, (1996).
3. South African Civil Aviation Authority. *ETOPS Operations*. South African, (2015).
4. L. Chester., M. Pandey., The New FAA ETOPS rule, Boeing Aero magazine, aero quarterly, Vol 2, (2007).
5. A. Osterwalder., Y. Pigneur., *Business Model Generation.* Jakarta: Elex Media Komputindo, (2012).
6. Morlok., K. Edward., *Pengantar Teknik dan Perencanaan Transportasi*, (Introduction to Transportation Planning Technique), Erlangga, Jakarta, (1988).
7. R. W. Boadway., The Welfare Foundations of Cost-Benefit Analysis*: The Economic Journal.* 84(336). 926-939, (1974).
8. A. K. Dasgupta. *Cost-Benefit Analysis. In: Economic Theory and the Developing Countries*. Palgrave: London, (1974).
9. F. Prasetya., *Modul Ekonomi Publik Bagian VI: Analisis Biaya dan Manfaat.* Malang: Universitas Brawijaya, (2012).
10. P. Holland., *Simple Introduction to Cost-Benefit Analysis.* Technical Support Services, Natural Resources Governance Unit, (2012).

Principle of generator HHO hybrid multistack type production technologies to increase HHO gas volume

Ajat Sudrajat[1,*], *Eva* Mayfa Handayani[1], *Noreffendy* Tamaldin[1], *Ahmad* Kamal Mat Yamin[1]
[1]Engineering Physics, Faculty of Engineering and Science, Universitas Nasional-Jakarta
JL. Sawo Manila No. 61, Pejaten, PasarMinggu, Jakarta Selatan 12520, Indonesia.
[2]GTriboE, Center of Advance Research and Energy CARe, Faculty of Mechanical Engineering, Universiti Teknikal Malaysia Melaka, Hang Tuah Jaya, 76100 Durian Tunggal, Melaka, Malaysia

Abstract. Hydrogen isclassified as New Energy and also considered as the most promising transportation fuel candidate in the future. Various pilots test of hydrogen fuel cell vehicles by the world's leading automotive industries since the last 50 years have begun to show bright spot in the utilization of hydrogen-based fuel cells as vehicle fuel. The electrolysis process of water (H_2O) would produce H_2 (hydrogen) and O_2 (oxygen). The conventional method resulted in inconsistent volume and quality of HHO gas. However, the current development of HHO gas production through electrolysis process varies in term of materials, production process, design of certain tools, and technical modifications to obtain optimum results. In this research, the Hybrid Multistack TypeHHO generator has been designed and developed by combining two types of dry and wet cell generators. In this study using both cell type generator (wet and dry cell) or called as a hybrid type. Through the process of electrolysis in HHO enclosure space, the HHO gas was produced. The volume of HHO gas obtained from the HHO generator as an alternative fuel is strongly influenced by the electrical current supplied and the concentration of KOH catalyst used. The test was conducted with four stages of catalyst amount from 5.6g/L; 11.2g/L; 16.8g/L; and 22.4g/L. The applied current is linearly increased, with theincreasing HHO gas production. It is proven when with the amount of catalyst used at 22.4g/L, the average HHO gas produced is 230.3mL/min. The author analyzes the performance of the generator in term of current and HHO gas production at a predetermined 12V constant voltage.
Keywords*-Cell generator HHO, HHO Gas, Hybrid Cell Generator, Calibration, Evaluation and optimization*

1. Introduction

The rapid development and technological innovation of transportation drives the need for fossil fuel oil demand. Its effects to global environmental issues demanding government to adopt policies on emissions generated from short, medium and long-term transport on land,

* Corresponding e-mail: ajat.sudrajat@civitas.unas.ac.id or ajatsudrajatr2@yahoo.com

sea and industry [1,9]. The issue of global warming and the depletion of fossil fuel reserves, encouraging the search for alternative fuels that are renewable, easy to obtain, easy to process, and expected to shift dependency on petroleum and promote environmentally friendly sources.

Energy utilization has been harvested from organic and non organic waste to become a useful source of energy for human. Utilization of water into electrical energy has been widely done with the presence of hydroelectric power. In addition to electricity, water can also be used as energy or fuel vehicles or stoves, through the process of electrolysis by converting water into HHO gas. HHO technology is still considered rare to do and developed, whereas this technology is very effective to suppress the use of fossil fuels. The basic materials for this technology is the water with abundant potential in tropical countries like Indonesia. [2]

HHO gas (Brown's Gas), is the result of electrolysis of water by using a direct electric current, thus splitting water into pure hydrogen and oxygen gaswhich has a high heating value. Until now HHO gas is used as an additional fuel in motor vehicles. By producing HHO gas as much as possible, it is expected to reduce the concern of the Indonesian people against scarcity and fuel price hike.

Until now electrolysis is the most widely usedprocess of producing hydrogen from water. Electrolysis is a process of decomposition of water molecules (H_2O) into hydrogen (H_2) and oxygen (O_2) with reaction-triggering energy in the form of electrical energy [9]. This process can take place when two electrodes are placed in water and direct current is passed between the two electrodes. Hydrogen forms on the cathode, while oxygen at the anode [11]

To increase the production of HHO gas produced from the electrolysis process, it is necessary to add the KOH catalyst dissolved in aquadest water electrolyte. This electrolyte as a catalyst that will reduce the energy required, so that the reaction rate for breaking water molecules become faster through chemical reactions that facilitate the process of decomposition of water into hydrogen and oxygen. It is because the catalyst ion can affect the stability of water molecules into H and OH ions which are more easily electrolysed. [11]

Research the production of oxygen gas through the process of electrolysis of seawater by using graphite as electrode and varying the electrolytes of NaCl and KOH. After research, the fact that the gas is produced in a salinity solution of 35% and a voltage of 13V is the oxygen gas and hydrogen gas. All of these studies produce oxygen gas and hydrogen gas. Production of measured and monitored oxygen is increasing as the voltage is increased. In addition, with increasing levels of electrolytes, increased oxygen gas production is also increasingly evident. KOH solution produces more oxygen gas than NaCl solution. The study states that the type of electrolyte and electric voltage affect the production of HHO gas. [10]

HHO generator is a tool that can convert water into hydrogen gas and oxygen. The addition of a HHO generator to a fuel-based engine can improve the combustion efficiency which means it can save fuel to produce the same mechanical energy. In this research has been designed and generated hybrid type HHO which combines two types of HHO cell type and wet cell and dry cell generator. To know the characteristics of the hybrid-type HHO generator, perform function tests and analyze the production volume capability or flow rate of HHO gas in units of milli liters per minute. The test results will be evaluated and validated to determine the performance of the HHO generator. HHO gas generated from HHO generators can be implemented on 1000CC engines up to 2000CC by injecting HHO gas through an air filter inlet without altering the engine's engine settings.

2. Hydrogen Gas Production

Hydrogen gas is known as Brown gas and a form of flammable hydrogen. Use of Brown gas is very broad, depending on the application. This gas can be produced from coal, steam reforming, and water electrolysis.

2.1. Hydrogen Gas from Coal

Coal is a natural wealth that is categorized as fossil energy that is formed from a very long metamorphosis process. The chemical structure of coal is by no means a simple carbon covalent chain. Optically coal is often a high-pitched chunk with varying water content.

2.2. Hydrogen gas from Steam Reforming

Steam Reforming is a method to produce hydrogen, carbon monoxide or other useful products from hydrocarbon fuels such as natural gas. This is achieved in a processing device called a reformer that reacts with steam at high temperatures with a natural material. Renewal of methane vapor is widely used in the industry to make hydrogen.

2.3. Hydrogen gas from Electrolysis

Gases generated from the electrolysis process of water are Hydrogen and Oxygen gas, with a composition of 2 Hydrogen atoms and 1 Oxygen atom. Electrolysis of water is an electrolysis process that is used to break water molecules (H_2O) into Hydrogen (H_2) and Oxygen (O_2). The process of electrolysis of water occurs with half the reaction of acid or alkaline (alkaline electrolysis) or both. In both types of reaction above, Hydrogen gas is also produced on negative electrode (cathode) and Oxygen gas is generated on positive electrode (anode).

The efficiency of electrolysis will increase when the production of hydrogen and oxygen gas is allowed to mix together so that the energy content increases as well. HHO gases should not be stored in high pressure tubes because these gases are highly explosive and can be burned 1000 times faster than gasoline vapor and automatically explode with heat around 570°C.

Electrolysis of pure water requires excess energy in the form of overvoltage to pass through the activation phase. Without excess energy, there will be no electrolysis at all and if it happens it will be very slow. Reactions that occur in the cathode and anode can be seen in Figure 1.

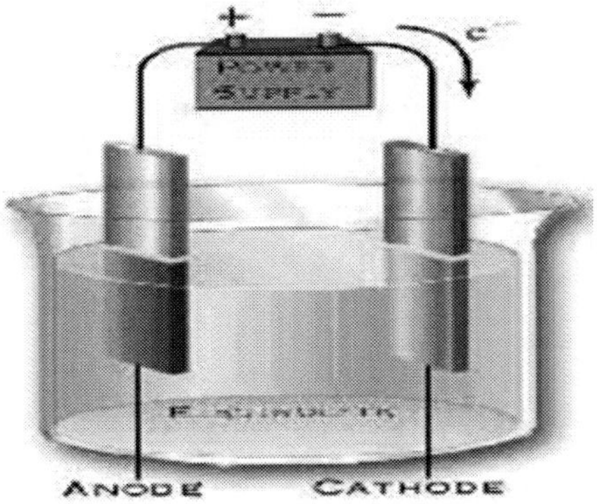

Fig. 1. Electrolysis process

reduction at cathode: $2\,H^+\,(aq) + 2e^- \rightarrow H_2\,(g)$
oxidation at anode: $2\,H_2O\,(l) \rightarrow O_2\,(g) + 4H^+\,(aq) + 4e^-$

3. HHO Generator Concept

The HHO generator is a device that can convert water molecules into HHO molecules. This change uses the concept of electrolysis to get the molecule. Electrolysis is a process of water decomposition (H_2O) into oxygen gas and hydrogen gas caused by the current passed to the water. DC resources are connected to two electrodes or two plates (usually made of inert metal such as platinum, stainless steel or iridium) which are then placed into water. The generator parts consist of:

3.1.Cell Generator

Cell generator serves as a place of electrolysis of separation of H_2O molecules to become HHO gas. This HHO cell generator have various parts that contribute to HHO gasproduction, including:

3.1.1 Electrode Plate

The electrode plate serves as electrical current conductor to the electrolytic water and the site for electrolysis. The electrode consists of anode and cathode plate. The material and extent of the electrode used affects the HHO gas generated from the waterelectrolysis process so that the electrode material must be selected from good electrical conductivity materials with corrosion resistance. Stainless steel type SS 316F, 316L, 316N, 317, 329, and 304 have excellent corrosion resistance in various environments, therefore are suitable as electrode in the waterelectrolysis process to produce HHO gas. In this study, the 316L stainless steel was used as electrode due to its low carbon content.

Electrode serves aselectrical current conductor from the voltagesource to the water to be electrolyzed. In electrolysis using DC current, the electrode is divided into two valves which are positive as anode and negative as cathode. This study utilized the 316L Stainless Steel type electrode plate and KOH electrolyte dissolved in distilled water. By dissolving the electrolyte in water it will increase the electrical conductivity. Electrolytes as catalysts in the electrolysis process can increase the reaction rate for breaking water molecules faster.

3.1.2 Spacer

Spacer is a series of nonconductive platesin between the stainless steel plate (SS316L) placed on the insulator material made of High Density Polyethylene (HDPE) with 3mm thick that serves as a barrier between the plate and as a leakage prevention electric current. The electrode plate is arranged in a 2x3 plat formation for each generator cell consisting of 6 cell HHO generators.

3.1.3 Gasket

The gasket serves as a barrier on the side portion of the HHO generator casing which serves as a water leak preventer for the gas from the HHO generator. The main requirement for this gasket should be able to close tightly between small gaps so that leakage from the side of the HHO generator frame does not occur. Materials used are rubber type MBR with 2mm thick.

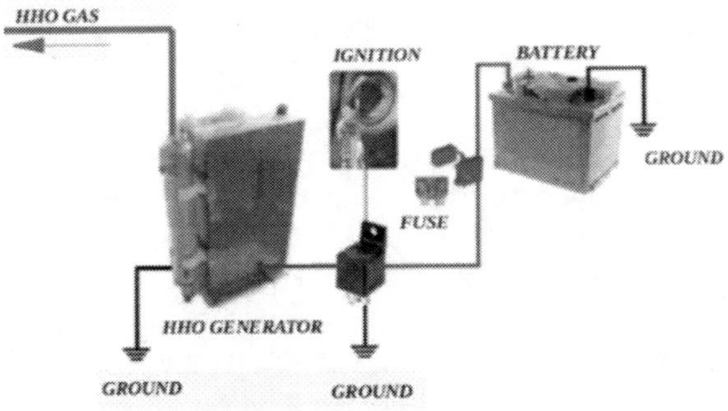

Fig. 2. Schematic diagram HHO generator

3.1.4 *Cell Generator Cover*

Cover used are from acrylic material. Cover serves as a cell cover generator (spacer) serves to clamp the arrangement of stainless steel plates. The plates on the left and right sides are mounted baud by welding as baud power serves as a current conductor to the anode (+) and cathode (-) electrodes.

3.1.5 *Connector*

The connector is a part that connects the outer and inner side which serves to fill the distilled water solution which has been mixed with the KOH electrolyte and the HHO gas exit as output. The connector is located on the top of the HHO generator on the left side for the input and the right side for the HHO gas output.

3.2. Type of Generator

3.2.1. *Dry Cell Type*

A dry cell type HHO generator, in which most of the electrodes are not immersed in electrolytes and the electrolyte only fills the gaps between the electrodes themselves. The advantages of a dry cell HHO generator are:

- Water fills the gap between the plat cells, the electrodes are not completely waterlogged.
- The heat generated is relatively small, because there is always a circulation between hot and cold water inside the HHO generator.
- The used electric current is smaller, because the converted power becomes less heat

3.2.2. *Wet Cell Type*

The wet cell-type HHO generator, where all the electrodes are submerged in the electrolyte fluid inside a water vessel.

The advantage of a wet cell HHO generator is:

- Generated gas is generally more stable and stable
- Generator maintenance is easier
- Design of making HHO generator easier

In the wet cell type, all areas of the plate electrodes are submerged in water for the electrolysis process to produce HHO gas.

3.2.3. Hybrid Type

Hybrid type generator is a combination of two types of HHO generator that is dry cell and wet cell. The hybrid generator has a formation where the dry cell generator is placed in a vessel containing the electrolyte liquid as in the wet cell type.

The advantages of hybrid type HHO generator are:

- The reservoir is present in a vessel containing an electrolyte water solution
- The electrode of the generator cell is immersed in the water of the electrolyte solution
- The heat generated is relatively small, because the water in the vessel can circulate well, without a water pump.
- The electric current used by the env is smaller, because the power converted becomes less heat
- No PWM electronics required because the working temperature is relatively low

Process flow diagrams are the stages or workflows in a hybrid-type HHO generator (see figue 2). The electrolysis process takes place at the HHO generator, where the water processing into the gas takes place in one vessel integrated in one place. The HHO gas output from the vessel is connected to the bubler tube through the top of the tube, then injected to the machine.

3.3. Catalyst

The catalyst is a material that serves to accelerate the reaction by lowering the activation energy and not changing the reaction equilibrium, and is very specific. The catalyst for water electrolysis uses a strong base electrolyte solution (KOH) to allow electricity to be easily transferred from one cell to another. The strong alkaline based electrolyte solution used is corrosive to metals similar to strong acids.

The concentration of catalyst (electrolyte) in water will affect the conductivity of the solution. The greater the volume of the electrolyte, resulted in greater conductivity of the catalyst molar, indicating that the ability of the solution to conduct electricity is greater or more easily flowing in the solution. The easier it flows at any time, then the solution can produce a larger electric current. Selection of KOH as an electrolyte because KOH easily absorbs water vapor. KOH has a high solubility in water that is 1100g / L.

3.4. Molecular Electrolyte Value.

Molality is the number of particles of solute (mol). Molality can be measured in solid form and can only be measured in mass, not its volume so that it is impossible to be expressed in the form of molarity. In this test, the catalyst used is Potassium Hydroxide or Potassium Hydroxide (KOH). The more catalyst dissolved, the greater the resulting production shown in the table below:

Table 1. Molasity of Potassium Hydroxide (KOH)

m = 0,1; KOH = 5,6g/L			m = 0,2; KOH = 11,2g/L		
V	C (A)	Produksi Hidrogen (ml/m)	V	C (A)	Produksi Hidrogen (ml/m)
12	2,4	100	12	3,2	175
12,5	2,9	125	12,5	3,9	225
13	3,6	175	13	4,7	250
13,5	4,3	225	13,5	5,5	300
14	5	300	14	6,3	325
m = 0,3; KOH =16,8g/L			m = 0,4; KOH = 22,4g/L		
V	C (A)	Produksi Hidrogen (ml/m)	V	C (A)	Produksi Hidrogen (ml/m)
12	3,9	200	12	4,4	250
12,5	4,7	250	12,5	5,3	300
13	5,7	300	13	6,3	375
13,5	6,8	350	13,5	7,3	425
14	7,8	400	14	8,2	450

4. Methodology

The steps taken in this study as follows:

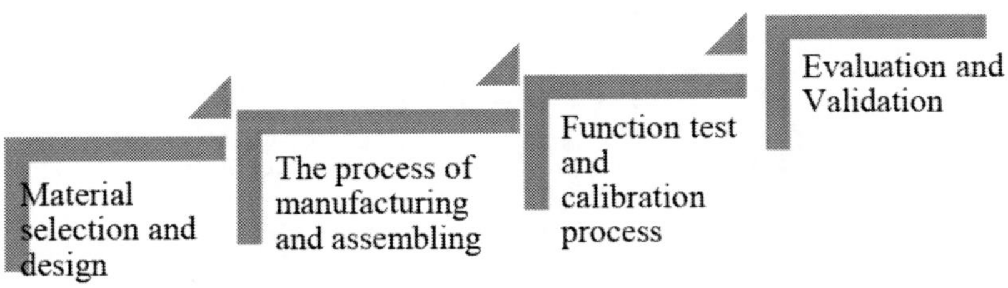

Fig. 3. Research Methodology Process

4.1 HHO Generator Manufacturing Process

Manufacturing process can be seen in the following flow diagram, where the work process starting from the design process, material preparation and testing are presented in detail.

4.2 MultistackCell Generator Design

HHO multistack generator design consists of 6 5x5cm stainless steel plate in each cell. The plates are arranged in parallel with the aim of obtaining more HHO gas production volumes with lower electric current intake than single stack models. To prevent electrical leakage, each plate is coated with a gasket. In the generator cells should be coated gaskets that serve as a barrier between the plate and as a leakage prevention electric current. The material used is HDPE with 1mm thick.

4.3 HHO Generator Case Design

HHO generator case is made from HDPE material with 80mm thick while cover is used from acrylic material. The cover acts as a cover or flank of stainless steel plates with surrounding bolts, and two bolts serve as a current conductor to the anode (+) and cathode (-) electrodes.

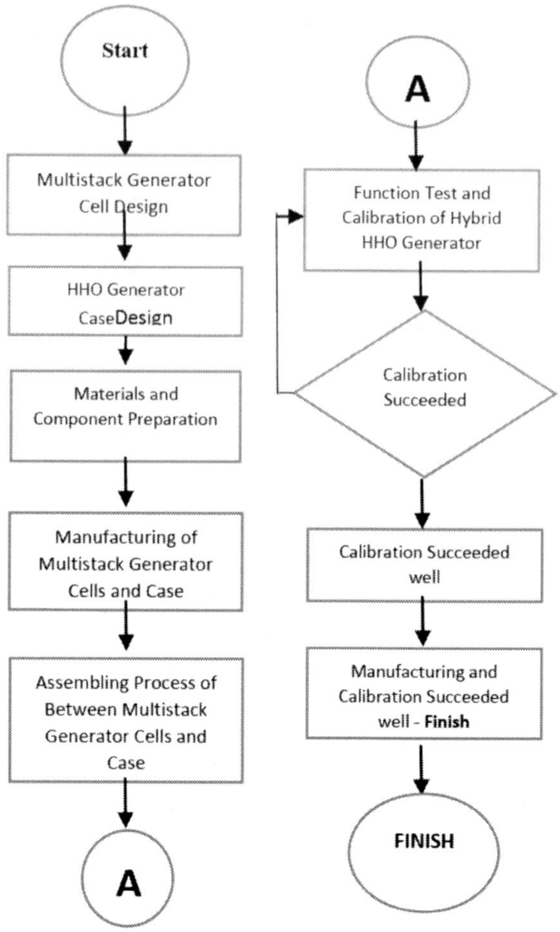

Figure 4. Flowchart of Manufacturing Process

4.4 Test Material Preparation

Materials to be used in designing HHO Generator, such as HDPE 80mm, HDPE 3mm, HDPE 1mm, 316L Stainless steel Plate, Amplas 1000, Acrylic Cuter, Bolt, Bubbler, NBR Rubber, Hose, and Niples.

5. Testing and Calibration

Before calibration of the HHO generator, first perform a function test on the device itself to determine the ability of the device in operation.

5.1 Function Test

The function test stage is performed to ensure that the HHO generator works properly and can know if there is leakage of electrolyte solution on the HHO generator case.

The next stage, carried out the calibration on the HHO generator in the following way:

Fig. 5. Testing and Calibration Process

5.2 Calibration Tool

The calibration tools used consist of Power Supply, ampere clamp, Flow meter and stopwatch.

5.3 Calibration Process

Calibration is done in the following way:

1. Turn on the HHO generator that has filled the catalyst solution with the first 5gram electrolyte and left for 10 minutes, then the HHO gas output through the bubler tube is measured using a gas flow meter.
2. Add 5gram electrolyte for every 10 minutes and measure the flow of HHO gas, until the number of electrolytes reaches 50gram.
3. This process is repeated three times in the same way.

5.4 Measurement Result Data

Measurement Data Result is the measured value of HHO gas output (liter per minute). This data is a large electric current relationship (Ampere) due to the addition of the amount of catalyst (gram) that is incorporated into the HHO generator vessel.

The calibration data is needed to know the accuracy of the HHO generator in generating HHO gas against standards stating the relation of electric current, the amount of electrolyte and the volume of HHO gas production.

Furthermore, the results of data processing compared with standard data as a reference that will be displayed in the form of tables and graphs. The data of calibration measurement can be seen in appendix 1.

6. Result and Discussion

6.1 HHO Generator Design Results

The result of HHO generator design is shown in Figure 6, is a series of HHO generator making process starting from the provision of ss316L plate material until assembling process becomes HHO generator cell. The next process combines the generator cell with the casing to become a functional HHO generator, shown in Fig. 6.

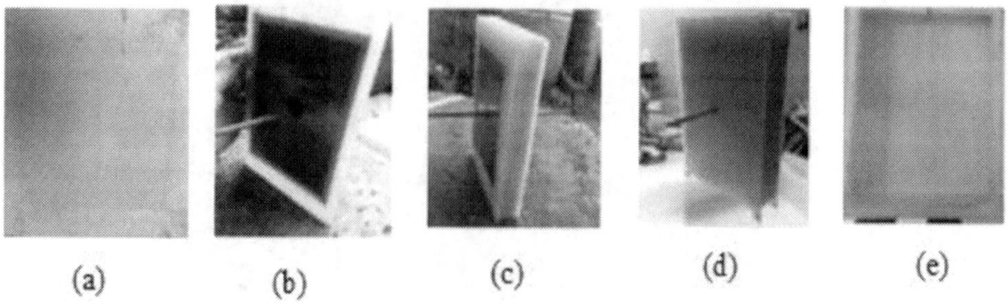

Figure 6. Asembling process of HHO generator cell; a. Pieces of SS316L plate, b. Preparation of Spacer (cell generator), c. Power Plate on Cell Generator, d. Cover Cell Generator, e. Frame generator HHO HDPE 80mm

After preparing the elements of the HHO generator, the next step is to combine these elements with the casing and complete it with the various accessories required. Figure 7 shows a hybrid-type HHO generator complete and ready to operate.

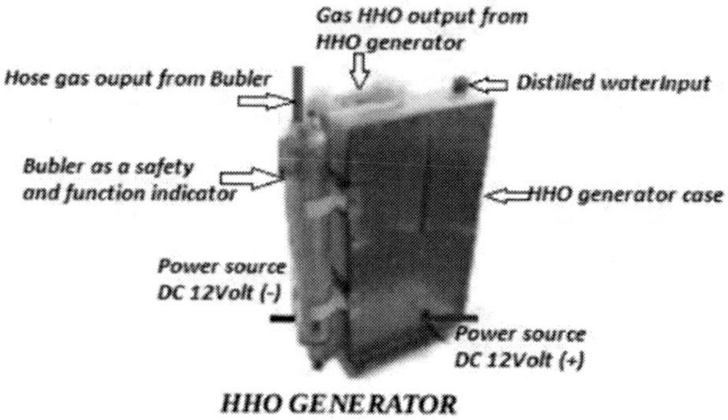

Figure 7. Generator HHO tipe Hybrid

6.2 Graph of Flow Relation and HHO Gas Production on Voltage

The following test results data on the HHO generator in the form of graphs. Testing done three times.

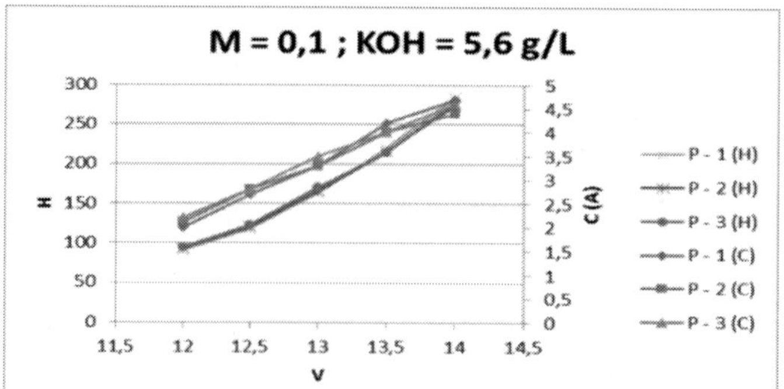

Figure 8. Graph current and gas production to voltage (M=0.1)

In the graphic Figure M = 0.1 shows the effect of voltage applied to the generator on the current and the production of gas produced with molarity =0.1; KOH catalyst content of 5.6gram/L. The graph shows the non-linear relationship to the standard on the number of 5.6gram/L catalysts caused by the unstable HHO generator because the electrolytic chemical process causing the change of resistance can affect the current change. Judging from the first test data up to the third test; the HHO generator tends to be more stable. When a voltage of 12V is applied, the current generated at the first test is 2A, the second test is 2.1A, and the third test is 2.2A, while the HHO gas produced at the first test is 92mL/menit, the second test is 94mL/menit, and the test third of 95mL /menit.

The other graphs in the second and third tests can be seen in Appendix 2 and the results in Table 2.

Table 2. Gas production on molarity to the resulting current

V	C (A)			
	M = 0,1	M = 0,2	M = 0,3	M = 0,4
	X1 (C)	X2 (C)	X3 (C)	X4 (C)
12	2,1	3,4	3,7	4,0
12,5	2,8	3,7	4,3	4,9
13	3,4	4,4	5,3	5,9
13,5	4,1	5,0	6,4	7,1
14	4,6	5,5	7,4	7,8

In Figure 9. Shows the average gas production generated by HHO generator from the three tests on each molarity, among others described in Table 3. Figure 10. Graph C (A) shows the average current generated by the HHO generator from all three tests on each molarity.

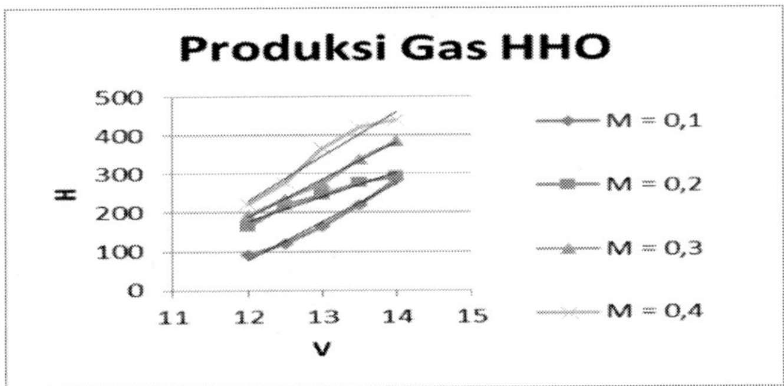

Figure 9 Gas production in varian molarity

Table 3. Molarity Catalyst on the volume of HHO gasproduced

V	Gas HHO			
	M = 0,1	M = 0,2	M = 0,3	M = 0,4
	X1 (HHO)	X2 (HHO)	X3 (HHO)	X4 (HHO)
12	93,7	167,3	193,3	230,3
12,5	121,0	217,0	236,0	279,7
13	167,0	238,0	278,7	358,7
13,5	216,7	277,0	332,0	414,7
14	280,0	292,3	380.0	434,7

Figure 10. Average electric current from all three measurements

The results of the current on the HHO generator test with 0.1 molarity large voltage 12V is 2.1A while the current according to the standard (literature) of 2.4A. Can be determined error by using the equation:

Measurement error against standard:

$$x = \frac{\text{HHO current} - \text{standard current}}{\text{HHO current}} \, x\, 100\%$$

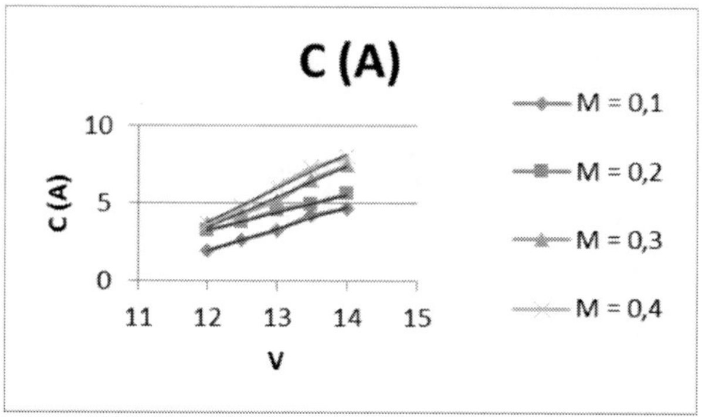

(1)

An error of reference value (default) with M = 0.1, voltage = 12V and current HHO = 2,1A, has a reference error of 0.14%.

The result of HHO gas production on HHO generator test with molarity 0.1 with 12V voltage is 93,7 mL / min while HHO gas production standard is 100 mL / min. Can be determined error by using the equation:

Standard value error:

$$x = \frac{\text{production of HHO gas} - \text{Standar of production of HHO gas}}{\text{production of HHO gas}} \, x\, 100\% \quad (2)$$

An error of standard value with M = 0.1 voltage = 12V and
HHO gas production of 93.7mL / min, has an error with a default value of 0.07%.

7. Conclusions

The results of this study can be summarized as follows:

1. HHO Type Hybrid Multistack generator was designed by combining two types of wet and dry cell type HHO generator shows good results with maximum error at current of 0.14% and on HHO gas production of 0.07% against standard.
2. With a constant voltage of 12V there is a different molarity variant to the electric current in the HHO generator. With a mean molarity of 0.1 the resulting current is 2.1A. For 0.2 molarity of 3.4A, 0.3 molarity of 3.7A, and 0.4 molarity by 4.0A.
3. The greater the molarity of the catalyst is given, the greater the HHO gas output rises at a constant voltage of 12Volt. The average molarity of 0.1 HHO gas

produced were 93.7mL/min, 0.2 molarity of 167.3mL/min, 0.3 molarity of 193.3mL/min, 0.4 molarity of 230.3mL/min.

4. For all tests, the error against the standard (literature) for each test on the resulting electric current and the resulting HHO gas is less than 1%.

In connection with the above results, the design of HHO multistack generator can reduce the electrical current and working temperature and provide accurate measurement value of HHO gas production volume of 1%.

Acknowledgement

In this research, the author would like to thank the Physics Engineering Laboratory of UniversitasNasional of Jakarta which has allowed the use of laboratory facilities as a place to conduct research and Faculty of Engineering and Science, UniversitasNasional, Jakarta, Indonesia (UNAS) for continuous support in this research.

We would also like to extend our gratitude to the Green Tribology and Engine Performance Research Group (GTriBoE), the Center for Advanced Energy Research (CARe), the Faculty of Mechanical Engineering, UTeM that has permitted the use of laboratory facilities.

References

1. **Harman, Arif. E, Hasan. D,** 2011, Effect of Addition of Oxyhydrogen Gas to the Performance of Gasoline-fueled Motor Pertamax, UniversitasHasanudin, Makassar.
2. **Sopandi. I, Hananto. Y, Rudiyanto. B,** 2015, Study of Thickness of Electrodes In HHO Gas Production (Hydrogen Hydrogen Oxygen) By HHO Generator Wet Type With NaHCO3 Catalyst (NatriumBikarbonat), Penerbit: ISSN 2085-2614, Jember.
3. **Rahmat. N. Fajri, Suryadimal, Arman. R,** 2014, Experimental Study on the Effect of Appearance of Oxyhidrogen (HHO) Gas on the Performance of Gasoline-Based Motorcycles Premium 88, Universitas Bung Hatta, Padang.
4. **Wahyudzin. I, Guntur. H. Laksana,** 2012, Study Characteristics of HHO Dry Cell Gas Generators and Their Applications on 1300cc Injection Engagement Vehicles, InstitutTeknologiSepuluhNopember, Surabaya.
5. **Satria. Y, Saragih. S. Abdi, Angin. N. Perangin,** 2012, Effect Analysis of HHO Dry Cell Generator Usage on Motorcycle Machine Performance, Universitas Islam Riau, Pekanbaru Riau.
6. **Ghiffari. Y. Arzaqa, Kawano. D. Sungkono,** 2013, Study Characteristics of HHO Gas Generator Type Dry Cell and Wet Cell Dimension 80x80mm with Addition of PWM E-3 FF (1 kHz), InstitutTeknologiSepuluhNopember, Surabaya.
7. **Arifin. T, Rudiyanto. B, Susmiati. Y,** 2015, Study of the Use of Neutral Stainless Steel 316 and Aluminum Electrode Platform on HHO Dry Cell Generator Performance,Penerbit : ISSN 2085-2614, Jember.
8. **Mulyono. S, Gunawan, Maryanti. B**, 2015, Influence of Use and Calculation of Premium Fuel Efficiency and Pertamax on Gasoline Gas Fuel Performance, Universitas Balikpapan, Pekanbaru Riau.

9. **Siamadi. M**, 2015, Effect of NaHCO3 Percentage Analysis on Production of Brown's Gas Results of Electrolysis of Water, UniversitasMuhammadiyahSidoharjo, Surabaya.
10. **Anindita. H**, 2011, Production of Oxygen Gas through Sea Water Electrolysis Process as Environmentally Friendly Source of Water, InstitutTeknologiSurabaya, Surabaya.
11. **Hidayatullah. P, Mustari. F**, 2008, The Secret of Fuel Water, Ufuk-Press, Jakarta.
12. **Wirawan. F, Kawano. DS**., 2013, The Effect of HHO Addition on Constant Diesel Engine Performance With KOH Catalyst Mass Variation In HHO Generator, InstitutTeknologi Surabaya, Surabaya.
13. **Salimy. D.H**, 2010, Production of Hydrogen Steam Reforming Dimethyl Ether (DME) Process with Low Nklir Temperature Reactor, JurnalPengembanganEnergiNuklir.
14. **Ailey**, 1996, Bailey's Industrial Oil and Fat Product, 5th Edition, Edited by **Y.H. Hui**, Wiley-Interscience Publication, USA.
15. **Damoko. D, Cheryan. M,** 2000. Kinetics of Palm Oil Transesterification in Batch Reactor, J. Am. Oil Chem, Soc. 77:1263-1237.
16. **Istiqomah. A. U, Rahmawati. F, Nugrahaningtyas. K. D,** 2016, Replacement of Fire Soda (NaOH) With Potassium Hydroxide (KOH) On Distillation of Water-Ethanol Binary System, JurnalPenelitian Kimia, Vol. 12 (2016), No. 2, Hal. 179-189, Surabaya.
17. **Figueiredo. J.L, Moulijn. J.A**, 2009, *Carbon and Coal* Gasification Science and Technology In. Proccedings of The NATO Advanced Study Instute On Carbon and Coal Gasification. Science and Technology Series: NATO Science, Vol. 105, p-672.
18. **Joshi. N, Naik. D,** 2015, Onboard Production of Hydrogen Gas for Power Generation, Journal of Research in Mechanical Engineering,Vol 2 - *Issue* 6 pp : 01-11, ISSN (*Online*) : 2321-8185, India.
19. **Shui. H, Cai. Z, Chunbao**, 2010, Recent Advances in Direct Coal Liquefaction, Anhui University of Technology, China.
20. **Robinson K.K**, 2009, Reaction Engineering of Direct Coal Liquefaction, USA.
21. **Jufang. H,** 2016, Design and Implementation of a High-Frequency Signal Generator Using the DDS Mixing Principle, Xinyu University, China.
22. **Pratama. R.A, Kawano. D.S**, 2013, Influence of Electric Frequency Usage To HHO Generator Performance and Engine Performance of Honda Karisma 125CC, JurnalTeknikPomits, Vol. 2, No. 2.
23. **Keche. A.J, Amba. G.P.R, Tated. R.G**, 2011, Efficient Utilization of Biomass By Gasification – A Case Study, International Journal of Advances In Thermal Sciences and Engineering, Vol. 2, Number – 1.
24. **Andre. P, Sekavenik. M, Mori. M**, 2011, Universal Model of a Biomass Gasifier for Different Syngas Compositions, University of Ljubljana, Slovenia.
25. **Silva. T.S.D, Senevirathne. L, Warnasooriya. T.D**, 2015, HHO Generator – An Approach to Increase Fuel Efficiency in Spark Ignition Engines, Sri Lanka Institute of Information Technology, Malabe, Sri Lanka.
26. **Murjito**, 2011, Design of Dry Cell Electrolyzer System for Automotive Fuel Saving, Universitas Muhammadiyah Malang, Malang.

27. **Akanksha, Chaurasia. P.B.L**, 2012, Home Gas by Hydrogen Fuel Cell, Suresh GyanVihar University, Jaipur, India, International Journal of Science and Research (IJRS).
28. **Bhavesh. V.C, Gaurav. P.R, Thusar. M.P**, 2016, An Experimental Investigation of HHO
29. Gas and Varying Compression Ratio on Performance Characteristics of Constant Speed Diesel Engine, IOSR Journal of Mechanical and Civil Engineering, ISSN: 2320-334X, Vol.13, Issue 2.

Mental workload of undergraduate student (a case study in Industrial Engineering Department of Diponegoro University)

Heru Prastawa[1], *Novie* Susanto[1,*], and *Try* Nofri[1]

[1]Industrial Engineering Department, Faculty of Engineering, Diponegoro University

Abstract. High competition in the era of globalization requires repair and improvement over all of the academic community in Indonesia, including the students. This makes the students compete in their studies and potentially cause mental studyload which could have implications for many things. This study aims to determine the level of mental load among the students of the Department of Industrial Engineering Undip in semesters 3 and 5 using the NASA-TLX, with respondents as many as 50 students. In addition, this study try to find out the NASA-TLX's dominant scale perceived by students and trying to prove, whether external factors such as previous semester grade point, study hours per day, entrance selection exam and free-time activities affect the score of mental load or not. The results of the mental study load measurement shows that the average score of mental study load of 80.04 (ponderable). For the dominant scale, the scale is Temporal Demand. Analysis of Variance (ANOVA) for external factors indicate that these factors do not have a significant influence on the existing score of the mental study load.

1 Introduction

Basically, human activity can be classified into physical work (muscle) and and mental work (brain). Although it can not be separated, it can still be distinguished between physically dominated work and work dominated by mental activity. This physical and mental activity cause some consequences, namely the emergence of workload. Workload can also be divided into two, namely the physical workload and mental workload.

Mental workload is defined by [1] as the difference between the workload demands of a task and the maximum capacity of a person's mental load in a motivated condition. A similar opinion is also put forward by [2] where the performance or working ability of a worker depends on the comparison between the size of the job demands and the amount of the worker's ability. If the task demand is greater than the employee's ability or capacity, it can cause overstress, fatigue, work accident, injury, pain, illness, and others. If the demands of the task is lower than the ability of workers, it can cause understress, boredom, boredom, and others. This comparison is then known as mental workload. Mental load measurements have long been known in the world of ergonomics, especially cognitive ergonomics. Since

* Corresponding author: nophie.susanto@gmail.com

the emergence of the NASA-TLX method in the 1980s, mental workload has become one of the factors calculated in various aspects relating to productivity and workers themselves.

Industrial Engineering Department as one of the departments at the Faculty of Engineering Diponegoro University also continues to make improvements that not only aims to improve the quality of graduates, but also aims to try to adjust the needs of the rapidly growing manufacturing industry. The preliminary study resulted in the following facts: 73% of respondents had difficulty arranging time between academic activities and activities outside of academics. As many as 69% of respondents have skipped lectures in pursuit of deadlines of laboratory reports or major tasks. As many as 63% of respondents suffered health problems so their daily activities are disrupted and even as many as 83% of respondents experiencing stress in undergoing lectures at this time. Through that preliminary studies, it can be concluded that the amount of academic burden that must be borne proportional to the mental burden that must be faced.

Some studies related this topic have been widely discussed as [4, 5, 6, 7, 8, 9, 10, 11, 12]. However, the specific topic related the mental study load of the student during the practicum and courses have never been discussed in detail. Based on this gap research and the backgrounf, it is necessary to conduct further research to determine the level of mental workload among students of Industrial Engineering Department of Diponegoro University. In addition, to find out the level of mental workload among students of the Department of Industrial Engineering Undip, this study find out and analyze the problems that are dominant perceived, as well. Besides, it find out whether the external factors such as grade point (indeks prestasi/IP) of previous semester, the number of study hours per day, entry point for a person to be a student of the Department of Industrial Engineering Undip until activities in his/her spare time affect the level of mental workload experienced by a person. In this study, NASA-TLX method [3] is used to complete the study aims.

2 Literature Review

The NASA-TLX method was developed by Hart of the NASA Ames Research Center and Staveland from San Jose State University in 1981 [3]. This method of was developed based on the emergence of subjective measurement needs that are easier but more sensitive on the measurement of workload. [3] describes the steps in measuring mental workload using the NASA-TLX method, ie:

1. Explanation of mental load indicator to be measured
2. Weighting
3. Rating calculation
4. Product calculation : ***Product = rating x Factor weight***
5. Calculation of Weighted Workload (WWL) : ***WWL = Σ Product***
6. Calculation of mental workload : $score = \frac{\Sigma(weight x rating)}{15}$
7. Interpretation of score

There are six indicator of mental workload in NASA-TLX: mental demand (MD), physical demand (PD), temporal demand (TD), performance (OP), Frustation level (FR), Effort (EF).

3 Method

The data collection for this research is performed by two stages, namely the design of questionnaire and the distribution. The criteria for this study object are undergraduate student of classes of 2014 and 2015. They are chosen because both of this generation is active undergraduate students who ever and is undergoing laboratory and project. The number of objects in this study are 50 students, with the provision of random sampling.

Dependent variable in this study is mental workload score, while the independent variables are previous grade point (IP) (2.75-3.00 ; 3.01-3.50 or ≥3.51), the study hour per day (≤1 hour, 1-2 hours, ≥2 hours), entrance point (SNMPTN, SBMPTN or UM), and spare time activities of the students (academic or non academic activities).

Data processing is done using ANOVA and NASA-TLX. The NASA-TLX analysis was performed using a fishbone diagram.

4 Result and Discussion

The identification of study load for student in third semester explain that there are about 20 credit (divided into 9 courses and 2 practicum). The study load for the student in fifth semester are about 18 credit (divided into 8 courses and 1 integrated practicum).

Based on the result of data collection, the average score of the mental workload using NASA-TLX is 80.04, while the total score for 50 respondents is 4002. According to the classification of [1], the score is classified as high scale and the work completed is quite heavy. Based on interviews conducted with the respondents, the most commonly reasons of this result related to some causes as seen in Figure 1.

The calculation of the number of each scale is intended to find the dominant scale of NASA-TLX which is felt by students of Industrial Engineering Undip. Details of the sum of each scale can be seen in table 1. From table 1 it can be concluded that the scale with the highest score is Temporal Demand scale and scale with the lowest score is Physical Demand scale.

Table 1. Calculation of NASA-TLX scale

No	Scale	Score	Rank
1	Mental Demand	110	4
2	Physical Demand	93	6
3	Temporal Demand	160	1
4	Effort	150	2
5	Performance	141	3
6	Frustation Level	96	5

The next step is analyzing the variance (ANOVA) for external factors with the result sd seen in Table 2.

5 Conclusion and Further Studies

The conclusion that can be taken from this study is the level of mental workload that occurred among students of Industrial Engineering Department of Diponegoro University semesters 3 and 5 is 80.04 (high level) high. The dominant scale is the Temporal Demand scale which is related to working time of a task or job as well as the strict deadline of the report.

Table 2 Result of ANOVA for external factors

No	External factor	Sub factor	Score range	Mean	Result
1	Previous grade point	2,75 – 3,00	66,00 – 89,33	83,65	Not significant
		3,01 – 3,50	44,00 – 92,00	76,85	
		≥ 3,50	55,33 – 90,67	76,30	
2	Study hour per day	≤ 1 hour	67,00 – 91,33	82,60	Not significant
		1 – 2 hour	44,00 – 94,67	77,85	
		≥ 2 hour	61,33 – 90,67	76,65	
3	Entrance selection exam	SNMPTN	48,67 – 94,00	78,40	Not significant
		SBMPTN	44,00 – 91,33	76,50	
		UM	66,00 – 94,67	84,95	
4	Free-time activities	Academic	58,67 – 94,00	85,15	Not significant
		Non academic	44,00 – 94,67	81,50	

The high score of temporal demand scale is caused by three reasons, namely the timeliness of reports and short tasks, the existence of a practicum with the timing of the concurrent execution and the habit of postponing the work owned by the respondent. For external factors such as previous grade point, study hours per day, entrance selection exam and free-time activities have not significant effect on the mental load score of existing measurement results. However, it does not rule out that the existing mental load score can be influenced by other external factors, which have not been tested statistically. The further studies related to the measurement of mental workload among students are:

1. The level of mental load based on other criteria, such as for students with the final report.
2. The level of mental load in each practicum module

References

1. Hart, S.G. *and* Staveland, L.E. *Development of NASA-TLX (Task Load Index): Result of empirical and theoretical research*. Amsterdam: Elsevier Science Publisher. (1988).
2. Tarwaka and Sudiajeng (2004),
3. Hancock, P.A. *and* Meshkati, N. *Human Mental Workload*. North Holland: Elsevier Science Publisher. (1988).
4. Astuty, S.M. Caecilia S.W. *and* Yuniar *Tingkat Beban Kerja Mental Masinis berdasarkan NASA-TLX (Task Load Index) di PT KAI Daop. II Bandung*. Jurnal Online Institut Teknologi Nasional. (2013).
5. Atma. *Pengaruh Jumlah Praktikum terhadap Nilai IP Semester Mahasiswa di Institut Teknologi Telkom.* Seminar Nasional Terpadu Teknik Industri Universitas Brawijaya (2013)
6. Budianto J. *Pengaruh Aktifitas Mahasiswa Dalam Berorganisasi dan Disiplin BelajarTerhadap Indeks Prestasi Mahasiswa Semester Ganjil Pada UKM Forum Pembinaan dan Pengkajian FKIP Unila.* digilib.unila.ac.id. (2014).
7. Kurata, Y.B, Lou, R.M. *and* Mathias A.C. Procedia Manufacturing. (2015).

https://doi.org/10.1051/shsconf/20184902017

8. Ramadhan, R. Tama, I.P. *and* Yanuar, R. Jurnal Rekayasa dan Manajemen Sistem Industri, Universitas Brawijaya 5 (II). pp 17-25 (2014).
9. Ramadhania, N. *and* Parwati, N. (2015). *Z.* jurnal.ftumj.ac.id. (2015)
10. Riyani, Y. Jurnal Ekos, 1 (VIII) pp 19 – 25. (2012).
11. Rubio. *Evaluation of Subjective Mental Workload: A Comparison of SWAT, NASA-TLX, and Workload Profile Methods*. Madrid: Blackwell Publishing Ltd (2004).
12. Rusyadi, S.H. *Hubungan Antara Manajemen Waktu Dengan Prestasi Belajar Pada Mahasiswa.* jurnal.fpsiums.ac.id. (2012)

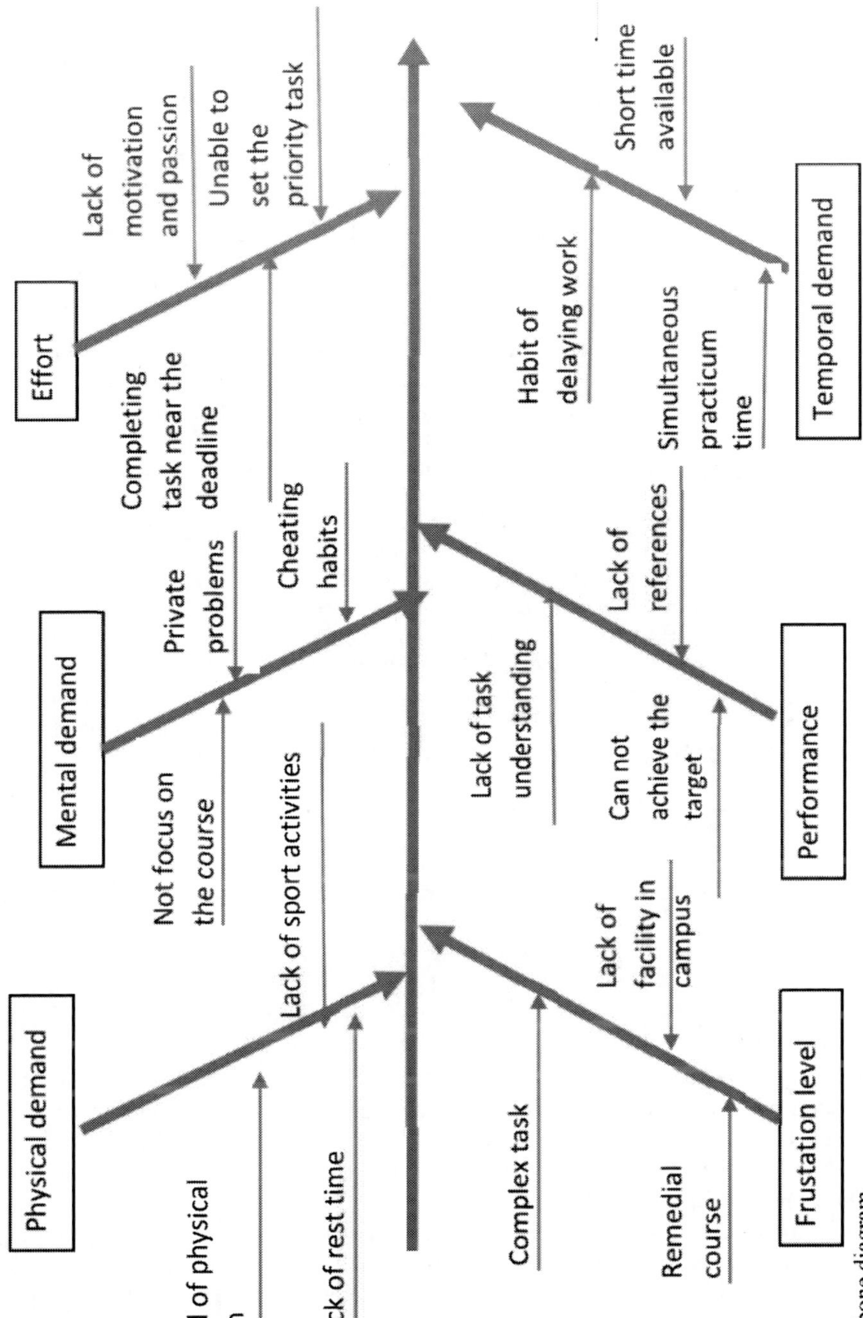

Fig 1 Fishbone diagram

Determination of copper metals in leachate using potentiometric method by concentration cells

Suheryanto[1,*], *Zainal* Fanani[1], *Dewi* Jayanti[1]
[1]Department of Chemistry, Faculty of Mathematic and Natural Sciences, Sriwijaya University

Abstract. The potentiometric method of cell type concentration is used to analyze the copper metal in leachate. This method is relatively good because it has a correlation coefficient of 0.9997 and linear region in the concentration range $1x10^{-5}$ - $1x10^{-3}$ mol L^{-1}. The precision of this potentiometric method is low because the persent of RSD produced is 0.6% with an average accuracy 0f 95.35%. The detection limit (LoD) of the measurable method is $1.085x10^{-6}$ mol L^{-1} with LoQ $1.313x10^{-6}$ mol L^{-1}. While the LoD instrument $1.217x10^{-6}$ mol L^{-1} and LoQ $1.560x10^{-6}$ mol L^{-1}. In addition, this method is very selective with the selectivity coefficient of -3.7 and it has uncertainty of measurement $9.7x10^{-7} \pm 1.282x10^{-6}$ to $9.12x10^{-5} \pm 3.5x10^{-5}$. Potentiometric method of cell type of concentration has an optimum condition at pH 5 and temperature 30°C.The conditions of copper in leachate water renged from $9.253x10^{-5}$ to $8.696x10^{-4}$ mol L^{-1}

Keywords: potentiometric, copper, leachate, concentration cell

1. Introduction

Leachate is a black waste water absorption containing organic and inorganic chemicals and a number of pathogenic or non-pathogenic bacteria that mutually decompose. Leachate water has the potential to contain heavy metals such as copper (Cu), lead (Pb), cadmium (Cd), mercury (Hg), arsenic (As), silver (Ag) etc. Copper is one of the heavy metals requires proper handling. Even in low-dose environments of about 1 mg/kg still have a toxic effect when inhaled in large amounts (3-4 mg/kg).

Several analytical methods such as atomic absorption spectrometry or AAS (Atomic Absorption Spectrophotometry), UV-Vis spectrometry and inductively coupled plasma (ICP) are used to define copper metals at low concentrations widely used in analyzing heavy metals in the environment is the potentiometric method. In addition, the method also requires complex time and samples.

Potentiometry is part of an electrochemical analysis technique, where there is a potential measurement of two electrodes at a current equal to zero. Potentiometric methods can determine the activity coefficients, pH values, dissociation constants and product solubility, standard affinity and chemical reactions. The advantage of this method is to have a wide range of concentrations (10^{-7} – 1 M); 0.1 – 5.0 % precision, easily essembled instrument,

* Corresponding author: suheryanto_mhg254@yahoo.com

low cost, adequate selectivity, low detection limit, high accuracy and can be used in color and cloudly solutions. In addition the potentiometric method allows the determination of heavy metals simultaneously between metals with each other.

In this research, the determination of copper metal in leachate water using potentiometric method based on cell concentration where a cell of two half cell consist of the same material but different ion concentration.

2. Recent Condition

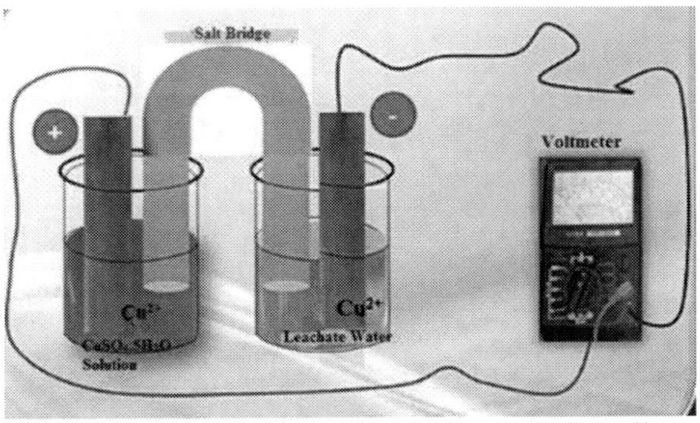

Fig. 1. Potentiometric circuits based on concentration cells

Analysis of copper metal in leachate water by potentiometric method using richmeter 102 to measure potential and copper metals as electrodes. All reagents were of analytical reagent grade. Demin water (0.9 μs conductivity) was used for all dilutions. The standard solution used for calibration is $CuSO_4.5H_2O$ were prepared by diluting stock solutions of 0.1 mol L^{-1} of each element.

2.1. Method Validation

Standard copper solutions that have been prepared with concentrations of 1.0 x 10^{-5} mol L^{-1} to 1.0 x 10^{-3} mol L^{-1} were measured potentiometrically by potentiometric method and made calibration curve and determined linear regression and linear equations to see slope and intersept values. The detection and quantification limits of the instrument are determined by measuring the potential blank of demin water as much as 10 repetitions. While for LoD method was determined by measuring the potential of standard solution with concentration of 1.0 x 10^{-5} mol L^{-1} to 1.0 x 10^{-3} mol L^{-1}. Determination of accuracy in analyzing copper metal with potentiometric method used spike method. Standard solution with concentration of 2x10^{-4} mol L^{-1}, 4x10^{-4} mol L^{-1} and 1x10^{-3} mol L^{-1} was added with leachate sample and potential measurement was done to the three spikes and calculated the value of accuracy using percent recovery. The precision of the method is determined by measuring the potential of a standard copper solution 6x10^{-5} mol L^{-1} concentration by 10 repetitions under the same conditions and time. Precision is determined by calculating the RSD value. The main ions (Cu) in $CuSO_4 5H_2O$ are made of three different concentrations of 1x10^{-5} mol L^{-1}, 1x10^{-4} mol L^{-1} and 1x10^{-3} mol L^{-1} mixed with foreign ions such as Cd^{2+}, K and Fe^{2+} with concentrations same. Then measured the potential of using multimeter and calculated the coefficient of selectivity.

2.2. Optimum Conditions of Analysis

To determine the optimum condition of a method, there are two variables in this research that is pH and temperature. To see the effect of pH on the electrode response, the potential was measured at a concentration of $1x10^{-4}$ mol L^{-1} with a pH of 3-8 and a scale of 1 each. The pH variations of (3-8) are made by adding nitric acid (HNO_3) or sodium hydroxide (NaOH). The effect of temperature on the potentiometric method was measured at a controlled by thermometer at different temperatures ranging from 20, 30, 40, 50 and 60°C.

3. Results and Discussions

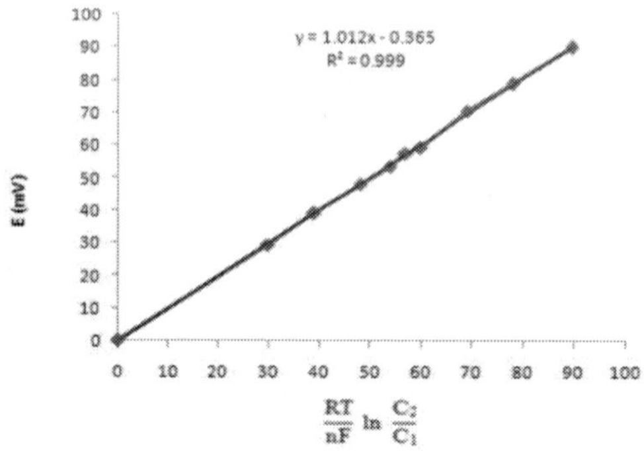

Fig. 2. Calibration curve of copper standard solution

Based on the calibration curve in Fig. 2, there is obtained a linear region in the concentration range of $1x10^{-5}$ to $1x10^{-5}$ mol L^{-1}. The slope value obtained from this study is 1.0125. This indicates that the slope is obtained in accordance with the set so that the method used has a good sensitivity. The detection limit is divide into two types: instrument detection limits and methods. Based on the calculation, obtained the instrument LoD value of $1.217x10^{-6}$ mol L^{-1} or 0.077 mg L^{-1} and LoQ $1.560x10^{-6}$ mol L^{-1} or 0.099 mg L^{-1}. In the LoD method obtained values of $1.085x10^{-6}$ mol L^{-1} or 0.069 mg L^{-1} and LoQ $1.313x10^{-6}$ mol L^{-1} or 0.083 mg L^{-1}. The precision value is 0.6 %. This indicates that the potentiometric method used in this study has good repeatability and accuracy. Accuracy value is obtained by calculating percent recovery. At concentration spike $2x10^{-4}$ mol L^{-1} obtained percent recovery 92.1 %, while for concentration $4x10^{-4}$ mol L^{-1} and $1x10^{-3}$ mol L^{-1} are 97.75% and 96.2 %. So the average percent recovery method of potentiometry based on cell concentration is 95,35 %. Percent recovery is not full 100 percent due to filtration process so that the components in the leachate sample and potential decrease.

Based on table 1, the calibration curve in pond 5 gives the largest contribution in the measurement uncertainty. Measurement uncertainty of copper metal in leachate ranges $9.12x10^{-5} \pm 6.7x10^{-6}$ mol L^{-1} to $9.7x10^{-7} \pm 8.5x10^{-7}$ mol L^{-1}.

Table 1. Contributors of uncertainty of measurement copper in leachate pools

Distributors	x	μ	μx/x
Weighing	6,3 gr	$5,774x10^{-5}$	$9,165x10^{-6}$
Pipette volume	1 mL	$4,057x10^{-3}$	$4,057x10^{-3}$
Volumetric flask	100 mL	0,068	$6,8x10^{-4}$
Volumetric	250 mL	0,126	$5,04x10^{-4}$

flask			
Precision		0,192	0,192
Acuraccy	95,35%	1,907	0,02
$C_{x(1)}$	55,3 mg L^{-1} or $8{,}696x10^{-4}$ mol L^{-1}	0,392	$7{,}089x10^{-3}$
$C_{s(1)}$	$9{,}12x10^{-5}$ mol L^{-1}		
$C_{x(2)}$	6,197 mg L^{-1} atau $9{,}751x10^{-5}$ mol L^{-1}	0,357	0,058
$C_{s(2)}$	$1{,}023x10^{-5}$ mol L^{-1}		
$C_{x(3)}$	3,995 mg L^{-1} atau $6{,}2866x10^{-5}$ mol/L	0,392	0,089
$C_{s(3)}$	$6{,}593x10^{-6}$ mol L^{-1}		
$C_{x(4)}$	2,852 mg L^{-1} atau $4{,}488x10^{-5}$ mol L^{-1}	0,356	0,125
$C_{s(4)}$	$4{,}71x10^{-6}$ mol L^{-1}		
$C_{x(5)}$	0,588 mg L^{-1} atau $9{,}253x10^{-6}$ mol L^{-1}	0,372	0,632
$C_{s(5)}$	$9{,}7x10^{-7}$ mol L^{-1}		

In determining the selectivity of a method, the effect of foreign ions on the main ions is used. Figure 3 illustrates the addition of foreign ions together such as Cd, K and Fe in the main solution of Cu in three different concentrations

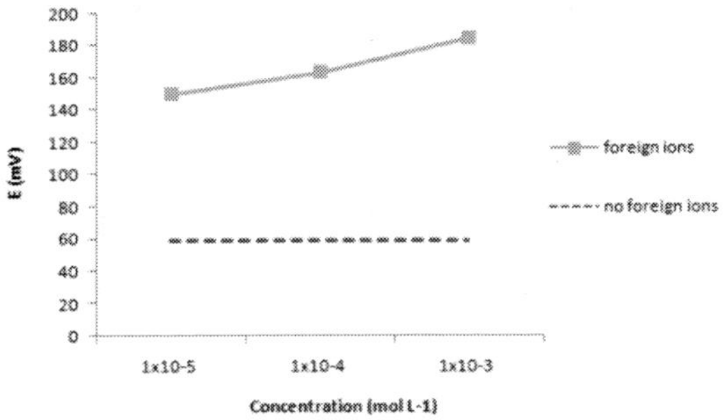

Fig. 3. The effect of the combined foreign ions on the cell potential

Based on the resulting graph, the potential value of the foreign ion is far from the potential value of copper in the absence of a foreign ions. If the concentration increases the greater the distortion generated by the discruptive ions. However, it does not affect the primary ion of copper because the resulting K_{ij} value <1 as described in table 1 below.

Table 2. $K_{i,j}$ values of combined foreign ions to main ions

Concentration (mol L^{-1})	$K_{i,j}$ Values
$1x10^{-5}$	-10
$1x10^{-4}$	-1
$1x10^{-3}$	-0.1

Different temperatures will affect the electrode used so that the slope value and ionic activity will be affected. A temperature of 30°C is an optimum condition because it is very close to the theoretical Nernst value. At a low temperature of 20°C it provides a potential price deviating from the theoretical Nernst value. This is because at a low temperature, the solubility is smaller so that the potential produced is not Nernstian. High temperature changes from 40-60°C can increase mobility or ion

 https://doi.org/10.1051/shsconf/20184902018

movement because the kinetic energy generated from the ion will increase so that the higher potential will be produced. Figure 4 shows the affect of temperature on potential.

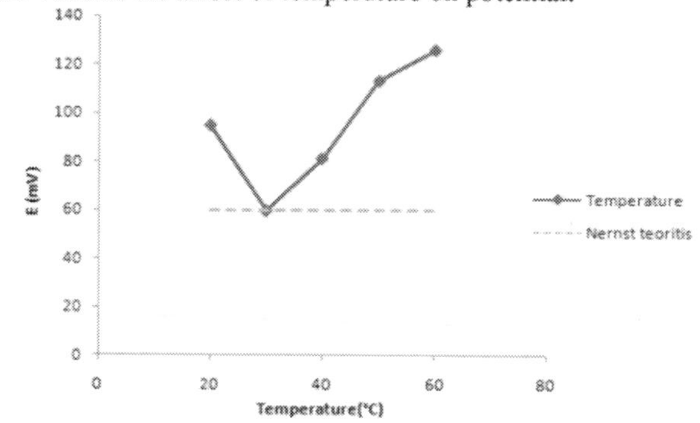

Fig. 4. Effect of temperature on cell potential

Figure 4 shows the effect of pH on the cell potential. At pH 3 and 4 the potential value is not so good, this is due to the presence of H^+ ions affecting the potential caused by Cu^{2+} so that its activity is not compatible in the test solution. The excess of H^+ ions will affect the equilibrium in a solution having a low pH where the equilibrium will shift so that the electode's ability to interact with Cu^{2+} ions is less and the ability to respond to Cu^{2+} ions.

The condition at pH is the optimum condition because the potential yield is close to the theoretical Nernst value. At pH 6-8 it shows the increasingly descending potential of the theoretical Nernst because of the excess of the OH^- ions resulting in hydrolysis. In addition, excess OH will result in precipitation of the metal to $Cu(OH)_2$, thus interfering with potential measurements of standard copper solutions. Figure 5 illustrates the effect of pH on the cell potential

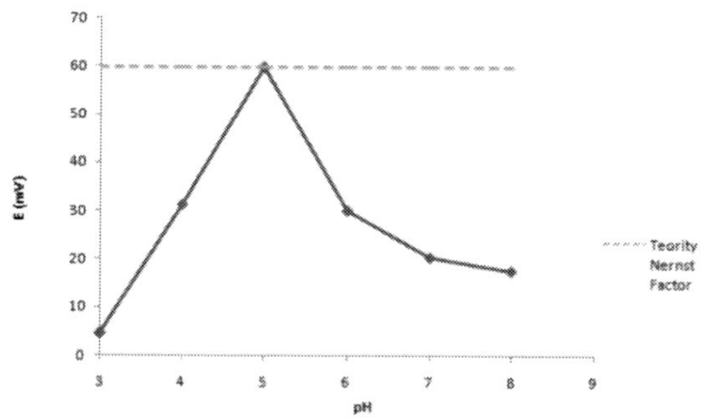

Fig. 5. Effect of pH on cell potential

The concentration of copper in leachate obtained ranged from 9.253×10^{-5} to 8.696×10^{-4} mol L^{-1}. Table 3 describes the levels of copper metal in leachate pools 1 to 5.

Table 3. Copper content in leachate water

No	Location	Cu (mol L^{-1})
1	The first pool	8.696×10^{-4}
2	The second pool	9.751×10^{-5}
3	The third pool	6.2866×10^{-5}
4	The fourth pool	4.488×10^{-5}

5	The fifth pool	$9.253x10^{-5}$

The first pool has the highest copper metal content because it is the first stream of leachate flow so that the components of the pool are still very complex. While in the fifth pool, the copper content in the place is the smallest because the metal content and other components have been reduced.

4. Conclusion

Potentiometric methods based on concentration cells used in analyzing copper metal in leachate have good validation results. In its use, the method has an optimum condition where the measuring limits can be used under certain conditions such as pH 5 and temperature 30°C. Potentiometric methods based on cell concentration were also successful in determining the concentration of leachate samples.

References

1. Anonim. (2009). *Way of Lead Test (Pb)by Atomic Absorption Spectrophotometry (SSA-Flame)*. Jakarta : National Standardization Agency.
2. Anonim. (2008). *Water and Wastewater - Section 59: Sampling Methods Wastewater.* Jakarta : National Standardization Agency.
3. Arias. (2015). *Electrochemical Cells Experiment 7*. Los Angeles: Los Angeles Harbor College.
4. Fardiyah, Q., Rumhayati, B., Rosemiyani, I. (2015). Effect of Ion Na^+, Na^+ Ion and Na^+ Ion Against Potential Correction Performance of Lead Ion (II) Coated Wire Type Based Pirofilit. *Journal of Chemical Valence: Journal of Research and Development,* **1**(2),80-83.
5. Fifield, F.W and Kealey, D. (2000). *Principles and Practice of Analytical Chemistry Fifth Edition*. United Kingdom : Blackwell Science Ltd.
6. Frag, E.Y.Z., Ali, T.A., Mohamed, G.G and Awad, Y.H.H. (2012). Construction of Different Types of Ion-Selective Electrodes, Characteristic Performances and Validation for Direct Potensiometric Determination of Orphenadrine Citrate. *International Journal of Electrochemical Science*, **7**(1),4443 - 4446.
7. Ganjali, M.R., Aghabalazadeh, S., Khoobi, M., Ramazani, A., Feroumadi, A., Shafiee, A and Norouzi, P. (2011). Nanocomposite Based Carbon Paste Electrode for Selective Analysis of Copper. *International Journal Electrochemical Science*, **6**(1),52-62.
8. Isa, I.M., Sohaimi, N.M., Hashim, N., Kamari, A and Mohamed, A. (2013). Determination of Salicylate Ion by Potentiometric Membrane Electrode Based on Zinc Aluminium Layered Double Hydroxides-4 (22,4-dichlorophenox) Butyrate Nanocomposites. *International Journal of Electrochemical Science*, **8**, 2112-2121.
9. Khani, H., Rofouei, M.K., Arab, P., Gupta, V.K and Vafaei, Z. (2010). Multi-Walled Carbon Nanotubes - Ionic Liquid - Carbon Paste Electrode as a Super Selectivity Sensor: Application to Potentiometric Monitoring of Mercury Ion (III). *Journal of Hazardous Materials*, **18**(3),402-409.
10. Koryta, J., Dvorak, J and Kavan, L. (1993). *Principles of Electrochemistry second Edition. Chicester* : John Wiley & Sons Ltd.

11. Purwanto, A., Supriyanto, C., and Samin, P.(2007). Test Validation of Cr, Cu, and Pb with Atomic Absorption Spectrometry Method. *Prosiding PPI-PDIPTN. BATAN*, 151-158.
12. Styarini, D. (2011). Validation of Chemical Analysis Methods. *News of Analytical Chemistry*, **19**,24-29.
13. Tandiayu, R. (2006). Optimization of PVC Based Membrane Composition with Ionofor N,N'-Dibensil-1,4,10,13-Tetra-Oksa-7,6-Diaza-Siklooktadekana for Making ESI-Cd (II). *Essay*. Makassar : Hasanuddin University
14. Thahir, S.Y., Atikah and Fardiyah, Q. (2013). Effect of pH and Temperature on The Rhodamin B Potentiometic Based Potential Sensor. *Chemistry Student Journal*, **1**(1),64.

Evaluation of advertisement existence toward attention and understanding level vehicle driver

Wiwik Budiawan[1,*], *Arfan* Bachtiar[1] , and *Veronica Desy* Afriyanti[1]

[1]Industrial Engineering, Department of Industrial Engineering, 50275 Diponegoro University, Indonesia

Abstract. When driving on the road many things can draw attention of the main task for driving, for example advertisement. The success of an advertising depends on how well it was able to communicate with the target properly. Information processing consisting of the perception process, organizing, and remembering of receiving information. This study tries to assess four-wheel drive behaviour when he saw advertisement on the road using The Eye Tracking method to know Allocation Time of Visual Spare Capacity, Attention Level and Understanding Level driver's. From 20 people who are respondents only 15 respondents have seen advertisement displayed and obtained research results that while driving, drivers have allocation of time of 6,52 second or 3,78 % of the whole long drive duration. With a small percentage, the attention of driver had low results seen from the perspective of eye integration for the advertisement. In this research also known that images and colour are easier to remember compared with the number of pictures displayed. So, the cognitive recommendation that can be given are the minimum distance between advertisement of 890 meters, Maximum element in advertisement contain 8 elements, Using the dark background in light coloured or otherwise, avoid several colours that hard to read if put together and Using Uppercase or Lowercase Font Style.

1 Introduction

Advertisement is defined as all forms of messages about anything that delivering through media, financed by the known initiator, and addressed to part or all of the community [1]. Type of advertisement are mass media advertisement include outdoor advertisement and internet, advertorial, Built In, posters, pamphlets and running text. Due to Outdoor Advertising Association of America, outdoor advertisement divided into 4 categories, such as Billboard, street furniture, transit and alternative media. Billboard advertisement is the most choice designed to introduce a brand name. Advertisement usually located on the road with crowd traffic. Nowadays, billboard advertisement using digital technology in static one like text or picture and dynamic one like video, running text, etc. [2]

Information Processing consisting of the perception process, organizing, and remember information (e.g.: reading the advertisements on the road). The information processing

* Corresponding author: wiwikbudiawan@ft.undip.ac.id

theory is about cognitive learning who describes processing, storage, and recalled back memories. On information processing, information received and stored on short-term, medium-term and long-term memory [3]. Cognitive processes are consisting of several stages such as sensation, attention, perception, think, memory and motivation.

The successful of advertising depend on how well it was able to communicate with information desire and attitude to the right person, at the right time, and at right expenses. From the information processing occurring in advertising, there is one of two things to be observed are attention levels and understanding level of advertisement target. Due to James [4], Attention defines as centralization of the mind, in the form of clear and sharp, into one of some object that simultaneous or from systematic thought .The essence of attention is vocalization, concentration and awareness. The attention is withdrawn from one thing to handle another thing effectively, and are a condition of the opposite to perplexed, daze, and unawares. While the definition of understanding is the quality of information that can be understood by consumers. The level of understanding of the individual can be increased when the information given right, clear and concise [5].

Advertising billboards on roads having target the road user as pedestrians, vehicle drivers and passengers in a vehicle. Time to read and to understand advertisement between pedestrians, the driver vehicles and passengers on it are different. Pedestrians have time to process information in advertisements longer than other because they can stop while read it. But the road users have a limited time since being in a on roads where only just glance it.

But many researchers have hypothesized that many advertisements have low involvement or not well understood by people [6]. It is because there are factors that influences people's desire to give attention in advertisements. That is factored culture as a culture the local environment and the social class; a factor social like the reference, family, and social status; a factor private age, sex, interest, the state of the economy, work, lifestyle, personality, and psychological factors as motivation, perception and trust.

Now many relevant information and research relate to allocation sight the driver in traffic, as eye tracking, visual occlusion, visual attention. etc. In many situations, driver have the time to look at targets that are not relevant for driving. This means that in a typical situation, the driver has the freedom to choose between several satisfactory information intake patterns, and there is time to glance at things that are not traffic relevancy. That visual behaviour, including the amount of "visual spare capacity", has been found to vary with driving task demand. With the visual spare capacity reserve indicate that the driver was able to doing side tasks besides whose main task to drive. In research of [7], there are two issues relating to attention while driving is targeted detection and monitor vehicle around. In research [8] estimated that a third of driver sight attention may be allocated upon an activity that is not related to the tasks driving. Based on research [9] said that driver also observes view on it non-critical like natural scenery or other things along the road. For example, in advertisements that can be pulled the attention of the main task of driving [10]. In this study, research only in static advertisement that is along the roads.

From the explanation above to be said in visual activities there are allocation of time that unconsciously done by eyes to look things that are not related to duty driving either in a vehicle and outside vehicles. One of that things that can be pulled the attention of the main task of driving is look billboards are located on the roadside. Besides pulling attention, the advertisement also must have their level of understanding whose contents can be easy to understand by advertisement target themselves. The allocation of time in his sight when doing the main task driving and also the goals of advertisement to pull attention and understanding can be used to design a research.

This study observes on four- wheeled driver behavior of a billboard advertising on the road with analyzing their level of attention and the level of understanding advertisement target, especially driver and also the best duration time that can be used advertising company based on available allocation time. This study uses Driving Simulator with City Car Driving Software that provides the two types of the driving environment that usually have advertisements on the road side. This study uses video data analysis to know the respondent visual process. This study aims to identify 3 aspects which is the duration time of Visual Spare Capacity, the level of attention, their level of understanding. The duration time of Visual Spare Capacity and the level of attention can be measured through eye tracking method from observation video data. For measuring the degree of understanding using questionnaire in the aftermath of the simulation. The result of data processing it may be regarded as the result of being more objective. So, that can be obtained the level of attention and the level of understanding of the driver when sees an advertisement and can recommend the best duration time that can be used by advertising company that make effective and efficient advertising.

2 Methods

This study is focused on the driver visual when he saw advertisement on the road. In this study, respondents will run driving simulator in the urban road conditions areas with certain route randomly. In every simulation road, there are various advertisements. The driving simulator set like real condition. The research was done in the laboratory. The simulator consists of the Logitech G27 with completeness the wheel, a pedal set, a transmission, a driver seat, CPU, sound systems, and 3 LCD TV. Software used in the simulation this is city car driving. This game is a driving simulation using approach real conditions.

1. **Variables**: In this research using independent variable, dependent variable and control variables. The control variables are type of driving activities, type of roads, distraction intensity, time and weather in the simulation and 7 advertisement that appear in each road. While independent variables are road contour, time distraction appears, type of road used the modern district (straight road) and the business district (many intersections). And the dependent variables are the visual spare capacity allocation time, level of attention and level of understanding the driver.
2. **Research Design**: From variables above, the study researchers designed to use by researchers to solve research problems. The research will be conducted by experiment within a subject that each respondent does one condition in two types of road. The simulation held in daylight and good weather with different route in every condition. Duration of one condition that have seven advertisements appear is 2-3 minutes. In the visual activity in this case, visual spare capacity time is measured by duration of driver looking away from the main task of driving until return focus on their work as seen from driving video recording. And for level of attention is measured from the point of view of ocular view. Level of understanding is measured using their level of understanding questionnaire which adjusted with research need. Into this research stage as the proportion of respondents that is used is who pass the criteria of respondents. First, respondents do an adaptation with driving simulator device. After that, respondents are obliged to wear recorder glasses in recording video mode. Respondents were required to do one condition that has a random road which on a modern district road or a business district road. Respondents had not told the purpose and objective of research for driving normally. After doing one route, respondents fill a questionnaire after research that contains four parts. The data was undertaken at random in showing each condition to avoid miss in all the condition. In addition, researcher has determined 5 routes consisting of 3 routes a modern road district and 2 routes business district.
3. **Respondent**: Respondents in this research are Industrial Engineering Department Diponegoro University students with the following criteria: having driving license,

has driving experience at least one year, good physical and mental health – normal sight do not wear glasses or contact lenses and have good understanding. This research is simple experimental research simple with strict control experiment. According to [11] the size of the sample used for experimental research with small sample size between 10 and 20 samples. In this research, retrieval of sampling technique is by purposive sampling with the total sample 20 people.

3. Result and Discussion

Data collection consists of video data and questionnaires. Through video recording obtained the direction of movement pupil eyes be integrated into the eyes. While from questionnaire data gathered awareness of the respondents will be advertising for advertisements, elements advertising viewed by respondents and their level of understanding the respondents each advertisement.

1. Visual Spare Capacity Allocation Time

First, to identify Visual Spare Capacity Allocation Time is important to note whether the respondents see an advertisement or not. Less than 50 percent of the number of respondents, while for advertisement first seen much as 55 % of the number of respondents. The average number of advertising ads seen by each respondent were 2.5 or means that each respondent only can see 2-3 advertising ads in one condition. From first advertisement until seventh advertising shows that the percentage of respondents that didn't see it were increasing. This occurred due to respondents were not aware of research objectives so it is more focused on driving than see an ad billboard.

To identify visual spare capacity allocation time, researchers used the questionnaire on 7 advertisement each route. In that questionnaire contains 3 parts are first part consist of question about respondent see an advertisement or not, the location and theme of the advertisement. The result of this question is right and wrong. The questionnaire could be divided into 4 categories: "See and Remember Right" category, 'See and Remember Wrong" category, "See and Do Not Remember category, "Do Not See" category. Four categories were based on the human cognitive processes starting of sensation obtained from an object of vision in the advertisement. Sensation obtained then would indirectly causing attention of human beings themselves. Sensation and attention in advertising become more important because advertising is considered successful if they convey information in it to the people who has seen. In this research got that 20 respondents who perform the conditions in research, only first advertisement that has the percentage people seen more than 50 percent of advertisement onward while the percentage that does not see higher than the see. To know the visual spare capacity allocation time using data from 15 respondents who advertisement in condition. The following are shown in Figure 1 a recapitulation of the respondents' answers that is divided into four categories:

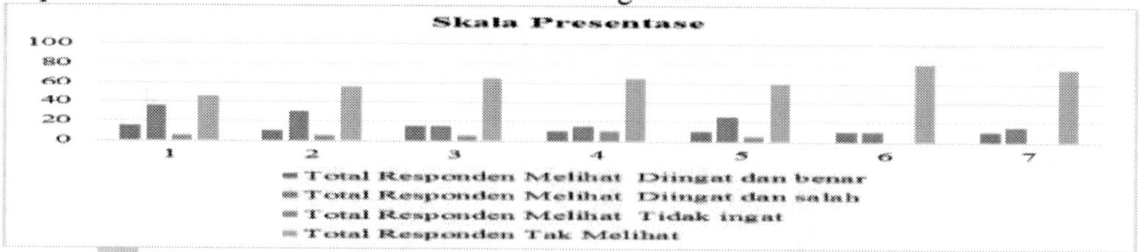

Number of Respondents See, Remember, and Answer Rightly
Number of Respondents See, Remember, and Answer Wrong
Number of Respondents See, Not Remember
Number of Respondents Not See

Fig. 1. Percentage of Respondent Awareness Chart

Here there are significant differences of from the percentage each category. Especially for advertisement banners is not visible to drivers. To respondents who see, the percentage respondents who see and remember wrong is higher than those who see and remember right.

Based on the "seen and remember right" category got that toilet equipment advertisement having the percentage largest by 34 % or 4 respondents, followed with an office rent and a flight advertisement with three respondents. This is caused toilet equipment advertisement display show detail image of toilet equipment such as a sink, a tap, a closet and type word "bathroom", so people can well aware of the ads clearly. For office rent advertisement showing a row of building and type the word "business loan", while a flight advertisement shows the images of planes and type the word "domestic flights" and "international flight" so that could deliver messages of advertisement. Visual spare capacity allocation time is the time it takes respondents when see an ad out the main task of driving. Visual spare capacity was calculated from two integration videos are video recording of Eye gaze application and video recording of recorder glasses when advertisement appear until it goes away. There is recapitulation of the following the average visual spare capacity time in each advertising of advertisement and percentage every condition of being displayed in table 1.

Duration driving among the respondents different because of the difference in speed and average mileage each route. The distance traveled in these simulations is 1,4 - 3,3 km and average speed 40 miles/hour. In these simulations, the average distance advertisement appears in the 350 meters. Average allocation time of visual spare capacity of 6,52 seconds with standard deviations 1,548 to take one advertising billboards with the percentage comparison with driving long duration as much 3,78 %. That can be expressed in performing a duty driving, the driver has an average allocation of time to see another one object driving in this task of billboard advertising for 6,519 seconds by standard deviations 1,548 % or by 3,778% while 1,5 - 6 minutes driving.

In the process see an advertisement, the visual scare capacity time longer indicates information processing longer, and oppositely. Timing is affected by advertising ads itself. For the longest time when respondents stop at traffic lights and vehicles stop. Time allocation also affected because the gap between advertisement appear every 350 meters that makes only a few more are viewed and remember correctly by respondents. Advertisement laid adjacent also affect information processing so many respondents were not remembered advertising seen.

2. Level of Attention

To identify attention of driver for advertisement appears can be seen from the direction of movement the eye pupil use Eye gaze application recording video who be integrated into the glasses recorder video that show visual area. Aside from the integration of two videos, the attention also be measured by the questionnaire. To make measure easy, both of video recording divided into several quadrants.

Based on the results two video recording integration and questionnaire obtained the result that of 15 respondents quadrant result of gaze direction different with the advertising quadrant ads appear. While from the 4 questions about its advertisement, shown suggests that images and color easier to remember compared with the letter and the number of images shown. As mentioned before, sensation and attention influential in communicating information in advertising. Through Eye gaze application video data to know the direction of the movement of the eye and video recording of regional sight wearing glasses recorder. A picture the results of recording video is divided into 9 quadrants to ease analysis like in Fig. 2.

Fig.2. Quadrant of Result Integrated Video

In addition, the result of the accuracy of the quadrant in the picture compares with third-part of questionnaire contain of the picture, the number of images, letter and color letter that shown worth right and not right.

Than 15 respondents who see an ad ads , only 2 respondents with each see 1st advertising with the results of the current positions of the advertising ads appear , the direction of motion of the pupil of the eye , and an answer with right .While as much as 23 different advertising are rested with the direction of the view so that not answer correctly , ad 14 are not in a quadrant From 15 respondents who see advertisement, only 2 respondents that can see an advertisement with result of advertisement appear location and eye gaze direction right. Then, 23 advertisements are located differently so respondent cannot answer correctly, 14 advertisements is not located in sight area, but respondent still can answer severe questions correctly and 13 advertisement that have a quadrant the current positions of the advertising element with a view but not answered incorrectly by respondents. At least inaccurate results the advertisement appear, the direction of movement eye pupil and answer caused respondents more focus on the main task of driving, elements in ads itself less attractive respondents' attention to see it. Based on research [12] said the driver will be realized objects in the located in the focus views or in front of vehicles. If an object outside the area, the driver see to the right and left horizontally. In addition, from the questionnaire obtained also that picture displayed and color letter easier to remember compared to the letter and the number of images shown. This suitable with research [13] that said an advertisement target spend more time to see the picture than a letter. Like in research [14] said advertising with a picture more interesting consumer attention than just a letter.

3. Level of Understanding

To identify the level of understanding of respondents to an advertisement seen by the duration time of the visual spare capacity and level attention respondent. In addition, to knowing the level of understanding the driver in advertisement using the fourth-part of the questionnaire. That passage consisting in four statements closed and one open statement. Four statements are relevant between images and letter that displayed, the use of images and letters that understandable easily, the use of interesting visual. To open statement, respondents answer about the meaning of advertisement are viewed. Open statement using Likert scale 1-5 (disagree; not agree; enough agree; agree; totally agree). Their level of understanding has only done in advertisement that appears in every route. This is because their level of understanding focused on the visual condition to advertisement. To identify their level of understanding, questionnaire forth-part gets the normality test. From that test got the data was not a normal distribution. From the Kruskal-Wallis test obtained from the result that there is no difference between each advertisement appeared from their level of understanding of respondents. From 4 statements are asked in the questionnaire and open statement, respondents was quite understanding about advertisement that they are seen.

https://doi.org/10.1051/shsconf/20184902019

4 Conclusion and Recommendation

The existence of advertising ads on a road providing an effect on the driver during a main task. While driving, drivers have 6,52 second allocation time or percentage allocation of time visual spare capacity or the time to see objects unrelated to traffic of 3,78 % of the whole long drive. The percentage is explaining in a duration driving, the proportion of drivers in seeing another object like advertising ads very small.

The attention when they saw advertising ads in driving visible through the direction of eye pupil movement and the eyesight area. From research getting a level of attention each person when saw advertising is low because many advertisements improper seen and improper answered by respondents. Elements advertisement shown also provides effect in attracting individual attention. In this study, picture displayed and color letter easier to remember compared to the letter and the number of the picture show.

Their level of understanding between advertisement has no significant differences that each person having a similar understandable toward advertisement who seen.

As the recommendations of the aspect of cognitive to the advertising agency who making advertising ads is the installation of advertising ads on the road more give attention at a distance more than 350 meters between advertising, so that the information distributed can be processed first. With the proportion of time very small, an advertisement should can use interesting elements advertising so can get the attention of people quickly of such as using pictures and color in advertisements. The use of language and the right word and adapted to the objective advertising can also help get attention. The use of visual spare capacity allocation time with the use of interesting and relevant elements advertisement can improve the level of understanding the driver toward advertisement. Thus, it can produce advertising billboards are effective, efficient and does not affect driving.

References

1. Anonym. Retrieved from Indonesian Advertising Company Association: http://p3i-pusat.com/definisi-iklan/ (2017)
2. M. Suyanto. *World Class Advertising Design Strategy (in Bahasa)*. Yogyakarta: Penerbit Andi (2006)
3. R. Slavin. *Educational Psychology: Theory and Practice Fourth Edition*. New Jersey: Neidenheights (1994)
4. W. James. *The Principles of Psychology*. **Vol.1**. (New York: Inc, 2007)
5. G. Tsoncheva. Measuring and Accessing The Quality and Usefulness of Accounting Information. *IZVESTIA - journal of University of Economics*. (Varna, 2012)
6. A. Mitchell, J. Russo, & M. Gardner. *Strategy-Induced Low Involvement Perception of Advertising. Working Paper, Graduate School of Industrial Administration*. (Carnegie-Mellon University, Pittsburgh, PA., 1979)
7. A. Borrowsky, D. Shinar, & Y. Parmet. Sign Location, Sign Recognition, and Driver Expectancies. *Transportation Research Part F : Traffic Psychology and Behaviour*, 459-465 (2008).
8. J.F. Antin, T.A. Dingus, & W.W. Wiewille. *An Evaluation of the effectiveness and efficiency of an automobile moving-map navigational display* (1990)
9. D.N. Lee, & M.F. Land. Where we look when we steer. *Nature*, 742-744. (1994)

https://doi.org/10.1051/shsconf/20184902019

10. J. Equist, T. Horberry, & I. Jhonston. Effect of advertising billboard during simulated driving. *Applied Ergonomics*, 619-626 (2011)
11. Roscoe. *Business Research Methods (in Bahasa)*. (Jakarta: Salemba Empat, 2006)
12. M. Crundall, E. Van Loon, & G. Underwood. Attraction and distraction of attention with roadside advertisement. *Accident Analysis and Prevention*, 671-677 (2006)
13. Q. Li, Z. Huang, & K. Christianson. Visual Attention toward Tourism Photographs with text: An Eye Tracking Study. *Tourism Management*, 243-258 (2016)
14. R. Pieters, & M. Wedel. Attention capture and transfer in advertising: Brand,Pictorial and Text Size Effect. *Journal of Marketing*, 36-50 (2011)

Risk analysis and strategic planning on bridge construction works at the toll road procurement projects in Central Java Province

Hery Suliantoro[1,*], *Nurul* Fitriani[2] and *Bagus* Hario Setiadji[2]

[1]Industrial Engineering, Diponegoro University, Semarang Indonesia
[2]Civil Engineering, Diponegoro University, Semarang Indonesia

Abstract. Risk is a condition caused by uncertainty. Risks will occur on any construction project, including bridge construction projects. Efforts that can be taken to minimize the impact of these risks are to engage in risk management activities. This research was conducted on bridge construction work on toll road procurement project in Pejagan-Pemalang, Pemalang-Batang and Salatiga-Kertasura. The purpose of this research is to analyze the risk of bridge development project in toll road project using Risk Breakdown Structure (RBS) method and then the result as database in discussing risk response strategy. The bridge construction project has 36 risks that are divided into six groups: materials and equipment, design, human resources, finance, management, nature and environmental conditions. Bad weather risks are the higest risk and seasonal risk causing temporary work stoppages. This risk-response strategy is avoidance. Short-term avoidance response strategy is to add shift workers, install tents and add additives in the acceleration of the process of maturation of concrete. The long-term avoidance response strategy is to evaluate and rearrange the work schedule by considering the weather forecast report.

1 Introduction

Some Toll Road Projects in Central Java Province which are parts of Mega Trans Java Project (Cikampek West Java - Gempol East Java along 652 Km) are still under construction. Among them are the Pejagan-Pemalang, Pemalang-Batang and Salatiga-Kertasura Road Projects. In those projects, there are bridge projects whose existence is quite essential; therefore it needs maximum handling efforts, including risk mitigation. Risk is a consequence of uncertain conditions in a project. There is a correlation between risk and the quality of construction project implementation [1].

Risks in bridge construction projects may include financial risk, external risk, design risk, management risk, construction risk, contractual risk, and health and safety risks [2]. Mitigation of risk can be carried out by conducting risk management plans. A risk management plan is considered as an attempt of analyzing risks that can enrich the decision-making process and provide additional arguments to help choosing the optimal

* Corresponding author: suliantoro_hery@yahoo.com

variant using multiple approaches [3]. Nevertheless, the risk management policies should be implemented and evaluated regularly to minimize the probability of failure [4].

This study aims to analyze the risks in the bridge construction project at the toll road project. In the analysis conducted will be known the risks that influence, value, category, and mitigation of each risk.

2 Literatur Review

2.1 Risk Management

The risks in the construction project are irreversible, but their impacts can be minimized [5]. An effort which can be done to minimize the impact is to carry out risk management activities [2]. International Standard (2009) [6] describes 5 (five) stages of risk management process: (1) Establish the context; This shows the relationship between the problem and the organization environment, the risk management process and the standard risk criteria, (2) Risk Identification; The identification process includes: what risks, when, where, how and why risks can occur, (3) Risk Analysis; Determine how much impact and possible risks will occur. Then the level of each risk is determined, (4) Risk Evaluation; Risk evaluation is carried out by comparing the estimated risk level against predetermined criteria and (5) Risk Treatment; Risk treatment is the stage of development and implementation of strategies and action plans for the selection of the best response.

2.2 Risk Breakdown Structure Method (RBS)

RBS is an aid for identifying, assessing, comparing, and reporting risk. A long list of risks can be hard to structure and prioritize, therefore the RBS can be a more effective tool to breaking down the risks and helping the project manager focus attention on certain risks. According to Hillson [7], RBS results risk response planning that will be useful as future project guideline.

RBS is a hierarchical structure of the source of risk, the method of grouping project risk based on the source that can organize and define the overall risk faced by a project. In general, RBS will be divided into 4 (four) levels, namely: References to printed journal articles should typically contain: Level 0 states a risky program, Level 1 is a global grouping such as a bridge construction project risk, Level 2 is a category of level 1 in the form of grouping of several risks such as planning risk group, risk contract group, operational risk group and so on, and Level 3 is a more specific risk description according to level 2.

3 Reseach Methods

The object of this research is the bridge project in the Pejagan-Pemalang Toll Road Project (17 bridges), Pemalang-Batang (27 bridges), and Salatiga-Kertasura (12 bridges).
Data collection method used in this research is observation and communication. Observation method used to get secondary data that is by searching data needed in book, journal, and other publication related to research theme. While communication method used to get primary data that is through interview and filling questioner.

Data collection begins with collecting secondary data to identify the initial risks which are then used as a reference for obtaining primary data through interviews and filling in questionnaires. Secondary data were taken using the observation method to obtain / identify the initial risks which were then used as reference in obtaining the primary data. Primary data is taken using the method of communication through interviews and filling questionnaires distributed in 2 (two) stages. Questionnaire I is a risk validation effort, questionnaire II is an effort to assess the frequency and impact level.

This research uses purposive sampling technique that is sample determination technique with certain considerations, so there is no limitation related to the number of samples. Respondents in this study are stakeholders that include contractors and supervisory consultants. Questionnaire I, respondents are experts / experts who have at least 5 years experience working on construction projects. While Questionnaire II, respondents are technical staff from contractors and supervisory consultants. Data analysis performed include nonparametric test, validity test, reliability test and risk analysis with Risk Breakdown Structure (RBS) method.

The purpose of this study was to explore the risk factors on the bridge project. This research is exploratory research. In this study, the factors affecting the bridge construction project are explored in depth. After obtaining the research factors with exploratory research, then these factors become the basis in preparing the questionnaire which is then analyzed to know the impact and its management. The method of analysis used is Risk Breakdown Structure (RBS). Hillson [7], states that RBS is a hierarchical structure of risk sources that can define the overall risk faced by a project. After the risk hierarchy is formulated then proceed with determining the value of the frequency level as a reference in categorizing and ranking each risk.

The variables in this study are risk factors derived from several references that are then validated using communication methods that is through questionnaires and interviews. The variables in this study can be seen in Table 1.

Table 1. The Exploration of Risk Factor

No.	Risk Factors	Reference
	Materials and Equipment	
1	Material Shortage	Asmarantaka, 2014 [8]
2	Material Error on form, function and specification	Asmarantaka, 2014 [8]
3	Delay of Material Supply	Asmarantaka, 2014 [8]
4	Material Damage in Storage	Asmarantaka, 2014 [8]
5	Material Scarcity	Asmarantaka, 2014 [8]
6	Inaccurate timing of material order	Asmarantaka, 2014 [8]
7	Delay of Equipment	Research Proposal
8	Equipment Damage	Winaktu et al., 2014 [9]
9	Equipment Shortage	Asmarantaka, 2014 [8]
10	Equipment Productivity	Asmarantaka, 2014 [8]
	Design	
11	Design Changes	Vidivelli et al., 2017 [10]
12	Design Variation	Zou et al., 2006 [11]
13	Error in Design by Planner	Asmarantaka, 2014 [8]
14	Error in drawing/specification	Taufik, 2010 [1]
15	Unclear information on the scope of work at the time of job description	Taufik, 2010 [1]
	Human Resource	
16	Less professional	Zou et al., 2006 [11]
17	Less labor	Zou et al., 2006 [11]
18	Low labor ability	Asmarantaka, 2014 [8]
19	Low labor productivity	Choudhry and Aslam,

No.	Risk Factors	Reference
		2011 [2]
	Finance	
20	Inflation	Choudhry and Aslam, 2011 [2]
21	Financial Failure of the Contractor	Choudhry and Aslam, 2011 [2]
22	Poor Estimation of Unexpected Cost	Research Proposal
23	Incomplete Cost Estimation	Zou et al., 2006 [11]
24	Unrealistic Cost estimation	Vidivelli et al., 2017 [10]
25	Late Payment	Vidivelli et al., 2017 [10]
	Management	
26	Error in Understanding Contract Document	Asmarantaka, 2014 [8]
27	Document is incomplete	Zou et al., 2006 [11]
28	Less coordination	Vidivelli et al., 2017 [10]
29	Dispute	Zou et al., 2006 [11]
30	Inadequate project information (soil test and survey report)	Zou et al., 2006 [11]
31	Error in choosing implementation method	Asmarantaka, 2014 [8]
32	Strict project schedule	Zou et al., 2006 [11]
33	Poor program scheduling	Zou et al., 2006 [11]
34	Delay during construction process	Vidivelli et al., 2017 [10]
35	Many work errors require rework	Research Proposal
36	Job implementation does not follow SOP	Asmarantaka, 2014 [8]
37	Bad K3 management	Asmarantaka, 2014 [8]
38	Poor surveillance and site management	Choudhry and Aslam, 2011 [2]
39	Subcontractor failure	Choudhry and Aslam, 2011 [2]
	Nature and Environment Conditions	
40	Earthquake	Winaktu et al., 2014 [9]
41	Landslide	Winaktu et al., 2014 [9]
42	Bad weather	Asmarantaka, 2014 [8]
43	Late permissions	Asmarantaka, 2014 [8]
44	A less secure project development environment	Asmarantaka, 2014 [8]
45	Disagreement interests with citizens	Research Proposal
46	Unstable government policy	Choudhry and Aslam, 2011 [2]
47	Excessive government approval procedures	Zou et al., 2006 [11]

Risk Probability and Impact Assessment is investigating the possibility of occurrence of some specific risks and the potential effects (negative impacts as well as positive impacts) of a project that may affect the project's ultimate objectives such as time, quality, price, scope of work. A commonly used analytical method is Probability and Impact Matrix, where risk can be arranged on a priority scale for the next step of quantitative analysis and risk mitigation.

Evaluation of the importance of each risk, as well as the priority of attention to risks can typically use Table 2. In this table the organization can determine anywhere from a combination of probabilities of risks and the effects of those riks in the classification of high risk, medium risk and low risk. The probability level values consist of values of 0.1 to 0.9 and an impact level value comprising values of 0.05 to 0.8. The probability and impact of the advantages and disadvantages can be analyzed in one matrix, by defining the good from each of the risks from different levels [12].

Table 2. Probability and Impact Matrix

Probability and Impact Matrix

Probability	Threats					Opportunities				
0.90	0.05	0.09	0.18	0.36	0.72	0.72	0.36	0.18	0.09	0.05
0.70	0.04	0.07	0.14	0.28	0.56	0.56	0.28	0.14	0.07	0.04
0.50	0.03	0.05	0.10	0.20	0.40	0.40	0.20	0.10	0.05	0.03
0.30	0.02	0.03	0.06	0.12	0.24	0.24	0.12	0.06	0.03	0.02
0.10	0.01	0.01	0.02	0.04	0.08	0.08	0.04	0.02	0.01	0.01
	0.05/ Very Low	0.10/ Low	0.20/ Moderate	0.40/ High	0.80/ Very High	0.80/ Very High	0.40/ High	0.20/ Moderate	0.10/ Low	0.05/ Very Low

Impact (numerical scale) on an objective (e.g., cost, time, scope or quality)

Each risk is rated on its probability of occurring and impact on an objective if it does occur. The organization's thresholds for low, moderate or high risks are shown in the matrix and determine whether the risk is scored as high, moderate or low for that objective.

Quantitative Risk Analysis is a risk analysis based on the risk priority resulting from qualitative analysis. In this method, the risk is analyzed to find out how much effect will be generated by giving a number that indicates the level of risk. The quantitative analysis is to assess the importance of risk to know which risks are most influential to the bridge construction project. The equation for calculating the value of risk level can be seen in equation 1.

$$\textit{Level of risk interest} = \textit{Probability x Impact} \quad (1)$$

This quantitative analysis should be performed again after the Risk Response Planning phase, as part of the Risk Monitoring and Control.

Nonparametric test was conducted to know the level of difference of understanding based on respondent data. The test was performed with the help of SPSS program in the form of testing two samples using Mann Whitney U Test. Validity test is a test used to indicate how accurately the instrument is used. While the reliability test is a test to determine the extent to which the consistency of the instrument used. The validity test is used to find out if there are any questions in the invalid questionnaire that should be discarded. While the reliability test is used to determine whether the instruments in the study can produce consistent data. The testing technique for validity test using SPSS software is by looking at the value of corrected item-total correlation. Testing technique for reliability test also use SPSS software by looking at coefficient value of alpha cronbach.

Once the ranking and risk categories are identified, then an analysis of the causes, impacts and respon of each of the risks can be done. Analysis of causes, impacts and risk management was obtained by conducting interviews with three experts. Risk response planning is a process of developing options and determining the most effective actions so that it is expected to increase the opportunity and reduce the risk seen from the negative side of the challenge. The type of response to risk is divided into four, there are : (1) avoid; the project team acts to eliminate risks or protect the project from its impact, (2) transfer; the project team transfers the impact of risk to a third party, (3) mitigate; the project team acts to reduce the likelihood of occurrence or impact of a risk, (4) accept; the project team decides to recognize the risks and does not take any action unless such risk occurs.

4 Result and Discussion

The analysis begins with identifying risk, followed by validity and reliability test. Data which is valid and reliable is continued to next analysis stage. Risks which affect the bridge construction project can be seen in Table 3.

Table 3. Risk Identification in Risk Breakdown Structure

Level 0		Level 1		Level 2		Level 3
Risky Program	A	Risks of Conducting Bridge Construction Project	1	Materials and Equipment	X1	Material Shortage
					X2	Material Error on form, function and specification
					X3	Delay of Material Supply
					X4	Material Damage in Storage
					X5	Material Scarcity
					X6	Inaccurate timing of material order
					X7	Delay of Equipment
					X8	Equipment Damage
					X9	Equipment Shortage
			2	Design	X10	Error in Design by Planner
					X11	Design Changes
			3	Human Resource	X12	Less professional
					X13	Less labor
					X14	Low labor ability
			4	Finance	X15	Inflation
					X16	Financial Failure of the Contractor
					X17	Poor Estimation of Unexpected Cost
					X18	Incomplete Cost Estimation
					X19	Late Payment
			5	Management	X20	Error in Understanding Contract Document
					X21	Document is incomplete
					X22	Less coordination
					X23	Dispute
					X24	Inadequate project information (soil test and survey report)
					X25	Error in choosing implementation method

Level 0		Level 1	Level 2		Level 3	
					X26	Less precise in applying the method of implementation
					X27	Poor program scheduling
					X28	Many work errors require rework
					X29	Job implementation does not follow SOP
					X30	Bad K3 management
					X31	Subcontractor failure
			6	Nature and Environment Conditions	X32	Landslide
					X33	Bad weather
					X34	Late permissions
					X35	A less secure project development environment
					X36	Disagreement interests with citizens
					X37	Unstable government policy

Validity test is performed on the frequency and impact of risk with corrected item-total correlation technique by comparing the r value of SPSS and r table (37 variables is 0.324). The variable declared invalid as the value of r arithmetic < r table is a design change with the value of r count frequency of 0.312. Reliability test performed on frequency and impact of risk with alpha cronbach technique. Based on the analysis that has been done, the instrument is stated reliable if the value of alpha cronbach> 0.6.

Recapitulation of respondent's answer to frequency and impact is analyzed and reciprocated to obtain value of risk level. Based on these values determined the category and handling strategies required.

Table 4. The Value of Frequency, Impact, Level of Interest, Category, and Strategy of Handling

Variable	Risks	Frequency (a)	Impact (b)	Level of Risk Interest (axb)	Risk Category	Risk Handling Strategy
	Material and Equipment					
X1	Material Shortage	0,346	0,231	0,080	Medium	Mitigation
X2	Material Error on form, function and specification	0,263	0,222	0,058	Low	Monitoring
X3	Delay of Material Supply	0,392	0,208	0,082	Medium	Mitigation
X4	Material Damage in Storage	0,346	0,216	0,075	Medium	Mitigation
X5	Material Scarcity	0,292	0,206	0,060	Medium	Mitigation
X6	Inaccurate timing of material order	0,354	0,194	0,069	Medium	Mitigation
X7	Delay of Equipment	0,413	0,255	0,105	Medium	Mitigation
X8	Equipment Damage	0,421	0,228	0,096	Medium	Mitigation
X9	Equipment Shortage	0,433	0,229	0,099	Medium	Mitigation

Variable	Risks	Frequency (a)	Impact (b)	Level of Risk Interest (axb)	Risk Category	Risk Handling Strategy
	Design					
X10	Error in Design by Planner	0,371	0,334	0,124	Medium	Mitigation
X11	Design Changes					
	Human Resource					
X12	Less professional	0,342	0,206	0,070	Medium	Mitigation
X13	Less labor	0,371	0,199	0,074	Medium	Mitigation
X14	Low labor ability	0,342	0,195	0,067	Medium	Mitigation
	Finance					
X15	Inflation	0,225	0,186	0,042	Low	Monitoring
X16	Financial Failure of the Contractor	0,238	0,271	0,064	Medium	Mitigation
X17	Poor Estimation of Unexpected Cost	0,379	0,175	0,066	Medium	Mitigation
X18	Incomplete Cost Estimation	0,358	0,175	0,063	Medium	Mitigation
X19	Late Payment	0,433	0,225	0,098	Medium	Mitigation
	Management					
X20	Error in Understanding Contract Document	0,233	0,193	0,045	Low	Monitoring
X21	Document is incomplete	0,350	0,184	0,065	Medium	Mitigation
X22	Less coordination	0,400	0,229	0,092	Medium	Mitigation
X23	Dispute	0,350	0,185	0,065	Medium	Mitigation
X24	Inadequate project information (soil test and survey report)	0,325	0,249	0,081	Medium	Mitigation
X25	Error in choosing implementation method	0,375	0,252	0,095	Medium	Mitigation
X26	Less precise in applying the method of implementation	0,358	0,274	0,098	Medium	Mitigation
X27	Poor program scheduling	0,400	0,206	0,083	Medium	Mitigation
X28	Many work errors require rework	0,346	0,255	0,088	Medium	Mitigation
X29	Job implementation does not follow SOP	0,354	0,297	0,105	Medium	Mitigation
X30	Bad K3 management	0,296	0,300	0,089	Medium	Mitigation
X31	Subcontractor failure	0,363	0,272	0,099	Medium	Mitigation
	Nature and Environment Condition					
X32	Landslide	0,363	0,327	0,119	Medium	Mitigation
X33	Bad weather	0,617	0,354	0,218	High	Avoiding
X34	Late permissions	0,421	0,220	0,092	Medium	Mitigation

Variable	Risks	Frequency (a)	Impact (b)	Level of Risk Interest (axb)	Risk Category	Risk Handling Strategy
X35	A less secure project development environment	0,358	0,238	0,085	Medium	Mitigation
X36	Disagreement interests with citizens	0,479	0,248	0,119	Medium	Mitigation
X37	Unstable government policy	0,304	0,218	0,066	Medium	Mitigation

Based on the analysis that has been done and presented in table 4, there are 1 (one) risk in high category that is bad weather risk with avoidance strategy, 32 risks in medium category with mitigation handling strategy and 3 (three) low category risk which are material error in form, function and specification, the risk of inflation and the risk of error in the understanding of contract documents with monitoring strategies.

Bad weather is a risk with the highest level of risk interest in the high category, followed by the risk of error in design by the planner which belongs to the medium category. Bad weather is a risk caused by natural factors, so it is often unavoidable. But the fact is, not all jobs are affected by this bad weather. In other words on certain jobs, there are many solutions to overcome these risks. The risks to the project can be eliminated and can even be protected against the impact of the bad weather. So in certain circumstances, this risk is categorized as a risk that can be avoided.

Bad weather risk causes the work to be stopped temporarily. Another impact is also the difficult transportation to enter the project site because the road becomes muddy and slippery. Handling efforts as short-term avoidance response strategies are addition of shift workers to pay back wasted work time. Regarding jobs which are not too dangerous and not at altitudes such as processing iron activity, it will be overcome by the installation of tents during the absence of strong winds. The addition of additives substance to accelerate the process of concrete maturation, supported by the use of non-conventional formwork which can be assembled and unloaded faster to pursue delay in progress of work. While the long-term avoidance response strategy are evaluation and rearrangement the work schedule considering the weather forecast report issued by Badan Meteorologi, Klimatologi, Geofisika (BMKG), so jobs affected by these risks such as casting work, girder beam installation etc. can be scheduled on a good weather. Despite being the highest risk, in fact this risk is a seasonal risk, because it only happens when the rainy season comes.

Another high-level risk is a design error by planners that impact on construction failures. A short-term mitigation response strategy is to find an independent expert design to conduct an evaluation. The project team evaluates and then discusses with the design engineering consultant to revise the design. A long-term mitigation response strategy is to conduct periodic coordination and review with design engineering consultant.

5 Conclusions and Suggestion

The results of the analysis can be drawn the following conclusions: There are 36 risks that affect the bridge construction procurement project. Obtained 1 (one) high category risk that is bad weather risk, 32 moderate category risk and 3 (three) low category risk that is material mistake on form, function and specification, inflation and error occurred in the understanding of contract documents. The risk of bad weather is the greatest risk and the seasonal risk, the handling effort are 2 (two) avoidance response strategies. Short-term response strategies are addition of shift workers to pay back wasted work time to replace the wasted work time, installaion of tents for work that is not at altitude, addition of additives substance to accelerate the process of concrete maturation supported by the use of non-conventional formwork. Long-term response strategy are evaluation and rearrangement the whole of project schedule by considering the weather forecast report issued by BMKG.

This research is conducted on a bridge project along a defined toll road project and does not classify on certain types of bridges. Therefore, it is advisable to conduct research related to risk analysis with more specific bridge classification.

References

1. Taufik, H.A.R.M, Pengelolaan Risiko Kualitas pada Tahap Pelaksanaan Konstruksi di Lingkungan PT. X, *Tesis* Universitas Indonesia, Indonesia, (2010)
2. Choudhry, R.M., and Aslam, M.A., Risk Analysis of Bridge Construction Projects in Pakistan, *Proceedings of International Council for Research and Innovation in Building and Construction (CIB) Working Commission W99.* Purdue University, (2011)
3. Dziadosz, A., and Rejment, M, Risk Analysis in Construction Project – Chosen Methods, Procedia Engineering 122, pp: 258-265, (2015)
4. C.S. Pawar, S.S. Jain, J.R. Patil, IJIRAE, **2**, 172-176, (2015)
5. A. Nurdiana, M.A. Wibowo, J.U.D. Hatmoko, *Procedia Engineering,* **125**, 12-17 (2015)
6. International Standard, *ISO 13000,* (2009)
7. D. Hillson, *Proceeding of, The Project Management Institute Annual Seminars & Symposium,* (2002)
8. Asmarantaka, N.S., Analisis Resiko yang Berpengaruh Terhadap Kinerja Proyek pada Pembangunan Hotel Batiqa Palembang, Jurnal Teknik Sipil dan Lingkungan, Vol. 2, no. 3, pp. 483-491A. Dziadosz, M. Rejment, *Procedia Engineering*, **122**, 258-265 (2015)
9. Winaktu, G., Mulyadi, L., and Hargono, E., Penentuan Skala Prioritas Risiko pada Pembangunan Jembatan Afiat Desa Kanigoro Kecamatan Pagelaran Kabupaten Malang, Jurnal Info Manajemen Proyek, pp. 11-19 (2014)
10. Vidivelli, B., Vidhyasagar, E., and Jayasudha, K, Risk Analysis in Bridge Construction Projects, International Journal of Innovative Research in Science, Engineering and Technology, Vol. 6 no. 5, pp 8271-8284 (2017)
11. Zou, P.X.W., Zhang, G., and Wang, J.Y, Indentifying Key Risk in Construction Projects: Life Cycle and Stakeholder Perspectives, *Proceedings of the 12th Pacific Rim Real Estate Society Conference*, Auckland, New Zealand, (2006)
12. Project Management Institute, *A Guide to the Project Management Body of Knowledge*, Fifth Edition PMBOK. Project Management Institute, Inc. Pennsylvania, (2013)

Quality management system design of vocational services in PM College based on ISO 9001:2015

Karina Gisela Napitupulu[1,*], *Mohammad* Mi'radj Isnaini[1], and *Dradjad* Irianto[1]

[1]Industrial Engineering, Institut Teknologi Bandung, Ganesa No.10, Bandung, Indonesia

Abstract. Vocational college institution has a substantial contribution to support the development of industrial human resources in Indonesia. PM College as one the reputable vocational college provides educational services for students and training services for industry employees. PM College has a focus to provide customer satisfaction. PM College's customers consist of industries mainly from manufacturing businesses. To guarantee the educational process, PM College has been applying the quality management system based on ISO 9001:2008. The current challenge of the PM College is to apply the latest ISO 9001: 2015 which is a revision of the previous version standard. ISO 9001: 2015 has some changes made as a form of adjustment. The institution feels the need to make improvements to the current quality management system to provide the satisfaction to customers, both from the industry and university students and to compete with the other polytechnic as well. This paper proposes an approach to prepare and analyze the organization in fulfilling the requirements of ISO 9001:2015. Several data processing and analysis were conducted to identify clauses' fulfillment between activity and document of the organization in meeting the ISO 9001:2015. The results of data processing and analysis show that PM College needs to make improvements in business processes and has not met ten clause requirements of ISO 9001:2015. Several adjustments were proposed to the business process and ten clauses to assist PM College in preparing the ISO 9001:2015 certification.

1 Introduction

The manufacturing industry is one of the driving force of industry in Indonesia. Polytechnics as educational institutions have a goal to produce industrial graduates who can work with professionals from the different field of expertise. As a manufacturer of human resources for the manufacturing industry, polytechnics are expected to provide the competent workforce. To produce a competent workforce, polytechnic need to pay attention and control the quality of their institution by applying quality management system.

PM College provides educational services for students and training services for industry and general employees. PM College has a focus to provide customer satisfaction.

* Corresponding author: iis@mail.ti.itb.ac.id

Customers of PM College education services consist of industry and business world who employ PM College graduates in their companies and students who receive PM College education services. PM College customer satisfaction is still below the level of interests and customer expectations indicates that PM College needs to strive to improve the quality of its services with ISO 9001 quality management system consistently. Therefore, it is necessary for PM College to improve the quality of education services and increase added value for customers who choose PM College by updating the current ISO 9001:2008 standard to the new ISO 9001:2015 to improve competitiveness among another polytechnic [1].

Moreover, the time allowed for the company to make the transition is three years after the latest standards are released. Such provision causes PM College to immediately update the ISO standard used by revising its quality management system to be recognized and re-certified. Therefore, PM College need to update the quality management system based on ISO 9001: 2015 to have guaranteed quality and can compete with competitors and answer customer expectation as education service user in PM College.

2 Theoretical Background

ISO is an independent international and non-governmental organization composed of 162 national standard bodies. According to ISO [2], the quality management system is a management system to direct and control the organization to meet the needs or defined expectations by paying attention to quality. International standards such as ISO 9000, ISO 9001 and other ISO related quality management standards encompass on several quality management principles: customer focus, leadership, people engagement, process approach, continuous improvement, relationship management and decision making with the factual approach.

The implementation of the quality management system is a strategic decision for the organization because it can help the organization to improve performance and design a basis for implementing sustainable development [3]. The organization may well-received higher benefits by implementing the quality management system to improve organizational consistency in providing products and services [4]. By meeting the customer requirements and improving customer satisfaction, the organization can identify risks and opportunities related to the context and objectives of the organization in demonstrating the ability to meet the standards [5].

ISO 9001 comprises the requirements that the organization needs to meet the standards of the quality management system by providing products and services to meet customer requirements. ISO 9001 covers different types of organizations with their respective activities and sizes. ISO 9001 has undergone several revisions from 1994, 2000, 2008 and up to date in 2015. At ISO 9001: 2015, there are three basic concepts used: process approach, PDCA cycle, and risk-based thinking. ISO 9001:2015 consist of 10 clauses namely: (1) Scope; (2) Normative Reference; (3) Terms and Definitions; (4) Organizational Context; (5) Leadership; (6) Planning; (7) Support; (8) Operational; (9) Performance Evaluation; 10. Improvements [6].

3 Proposed Methodology

The proposed stages to design and analyze the existing quality management system on PM College are depicted in Figure 1. The literature study was conducted by reading research, articles, journals, and books related to the quality management system issues especially ISO 9001: 2008 and ISO 9001: 2015. Data collection is conducted on the quality management system that the organization currently implements is ISO 9001: 2008. The data collected in the form of primary data and secondary data from the organization. Primary data collection conducted is an interview with stakeholders about the internal and external conditions of the organization, business processes and the implementation of quality management systems so far in the organization. While secondary data is obtained from organizational documents such as organizational profiles, organizational structures, and quality management system documents related to the application of ISO 9001: 2008.

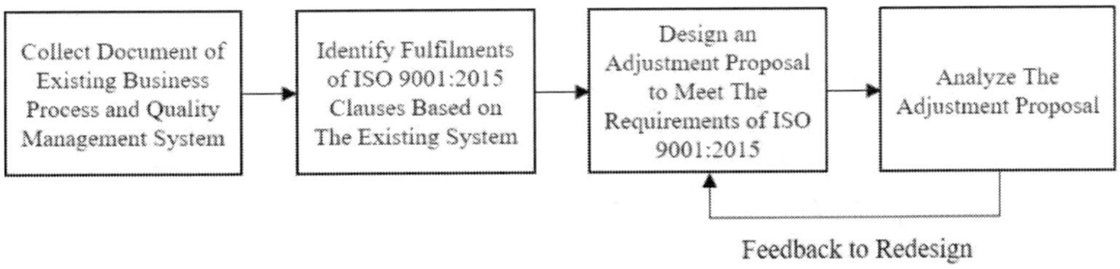

Figure 1. The proposed methodology to adjust the current system to meet the standard of ISO 9001:2015

Based on the data collected, data processing is performed by identifying the requirements of ISO 9001: 2015 clauses that have not been met through business processes and quality management system documents. It then identifies by checking the achievement to meet the requirements in each clause of ISO 9001: 2015. The results of clause identification indicate business process activities that have not met the ISO 9001: 2015 clause and documentation that has not been reached. The results of identification form the basis for designing documents to support compliance with ISO 9001: 2015 requirements. The design of the proposed quality management system is conducted by designing and improving the current quality management system documents. The thing done is the design of the proposed quality guidelines as one of the documented information that should be owned by every organization that implements ISO 9001.

The analysis stage includes an analysis of the organization's business process adjustments to meet the requirements contained in ISO 9001: 2015, an analysis of compliance with ISO 9001: 2015 clauses based on actual organizational conditions and design results, and an analysis of compliance with documentation required under ISO 9001: 2015. Lastly performed implementation analysis and potential barriers which may arise in the implementation of the quality management system in the organization. In the analysis process, validation is performed to determine whether the designed solution meets the requirements of ISO 9001: 2015 clauses and is relevant to the actual condition of the organization. The validation activity is done to discuss the result of the problem owner design that is PM College Quality Management Team.

4 Analysis

4.1 Business process categorization based on ISO 9001 :2015

The business process shows activities in an interconnected organization to produce a product or service. Business process mapping is carried out to meet the requirements of ISO 9001:2015 relating to the process approach. Every process found in the organization needs to be addressed to meet customer satisfaction. Figure 2 shows the business process which applies to the existing services in PM College today.

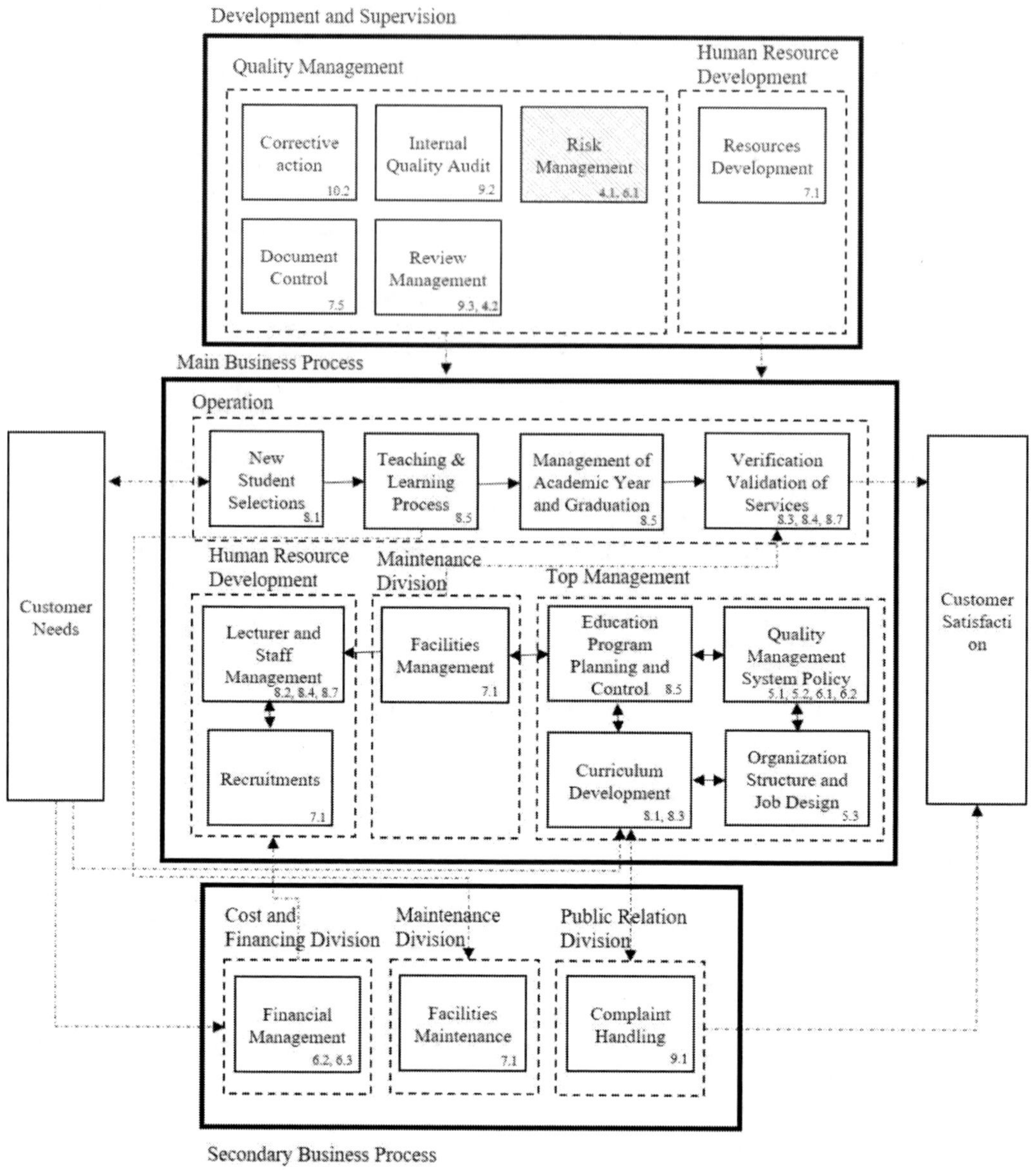

Figure 2. The existing of PM College business processes.

By analyzing the business process linkage with the clause of ISO 9001: 2015, the following clause are not yet available: Clause 4.1 on the understanding of the organization and its context; Clause 4.2 on understanding the needs and expectations of the parties concerned; Clause 6.1 on measures to deal with risks and opportunities, and; Clause 7.1.6 concerning organizational knowledge. The related activities of clauses 4.1 and 6.1 have not been implemented in the quality management system therefore it is necessary to add activities to the company's business processes. The related activities of clauses 4.1 and 6.1

were added to the Quality Management Representative section in the form of risk management. The result of business process improvement is depicted also in Figure 2.

4.2 Fulfillment of ISO 9001:2015

To find out the readiness of PM College to fulfill ISO 9001:2015, the analysis regarding the conformity of the management of the current quality management system with the clause requirements at ISO 9001: 2015 was performed. Table 1 describes the result of a comprehensive analysis and recommendation on the current quality management system.

Table 1. Unidentified clause fulfillment.

No	Clause	Recommendation
4.1	Understanding the Organization and its Context	Repair manual quality
		Proposed document form external and internal issue identification
4.2	Understand the needs and expectations of the interested parties	Repair manual quality
		Proposed documents of identification form of interested parties and requirements
4.3	Determine the scope of the quality management system	Quality manual adjustment
4.4	Quality management systems and processes	Improved document input process output
5.1.2	Customer focus	Repair manual quality
6.1	Measures to deal with risks and opportunities	Design of risk identification form
		Addition of risk matrix form
		The design of risk management procedures
7.1.1	Resource	Repair manual quality
7.1.6	Organizational Knowledge	The design of organizational knowledge procedures
		Designing an identification form for organizational knowledge needs
9.3.2	Management review	Quality manual adjustment
		The design of management review procedures

5 Analysis and Discussion

5.1 Business process categorization based on ISO 9001 :2015

In the application of ISO 9001: 2015, an understanding of the organization's business processes is required. The known business process is needed because ISO 9001: 2015 applies a process-based and risk-based thinking approach to every process undertaken to provide products or services. Based on the observations made to understand the business process of PM College that has been certified ISO 9001: 2008, there is no risk management activity because this activity is not required on the requirements of ISO 9001: 2008. However, risk management activities are required requirements in ISO 9001: 2015 so organizations need to add these activities to business processes. The addition to this business process shows that PM College runs several activities related to risk management in the processes and quality management system of the company.

The addition of new risk management activities is suggested in the process of development and management of business processes PM College. According to ISO [7], risk management activities are undertaken so that the organization can determine the factors that have the potential to cause the organization's quality management processes and systems to deviate from expected results and take precautions to mitigate impacts and maximize emerging opportunities.

5.2 Design implementation of the existing quality management system

Implementation of ISO 9001: 2015 raises some impacts on the organization. The impact of implementation is divided into two: administrative and organizational impacts. The administrative compound is related to changes to the applicable standards, procedures, work processes and policies in the organization. Organizational impact on the organization as an example of the addition of personnel and changes to the organizational structure. For the impact of proposed improvements from administrative impacts, adjustments made to PM College need to make changes to the company documents that is the creation of new procedures, improvements to the quality manual and document improvements shown in Table 2.

Table 2. Improvements on unidentified clause

Type of proposed improvement	No	Clause
Creating a new procedure	1	6.1 Measures to deal with risks and opportunities
	2	7.1.6 Organizational knowledge
Repair manual quality	3	4.3 Establish the scope of the quality management system
	4	5.1.1 Leadership and commitment
	5	5.1.2 Focus on the customer
	6	7.1 Provision of resources
	7	9.3.2 Input management review
Document repair	8	4.1 Understand the organization and its context
	9	4.2 Understand the needs and expectations of interested parties
	10	4.4 Quality management systems and processes
	11	9.3.2 Input management review

The results of the proposed document improvement are discussed together with the Quality Management Team as the party of PM College and evaluated and approved by the PM College regarding document aspect. Furthermore, the organization needs to verify the proposed documents to know whether the documents are designed as needed and not redundant with other documents. Implementation of verification activities may be performed by the head and staff of the relevant sections. If the document has passed the verification stage, numbering on the proposed new document and the new document register in the document list can be made. In addition to the preparation of new documents that need to be done, the company also needs to adjust the quality manual of the company.

6 Conclusion

The design of quality management system that is done on education service at Bandung State Manufacturing Polytechnic done by renewing the previous quality management system which refers to ISO 9001: 2008 standard. The design is done by adjusting business processes and adjusting the requirements. Adjustment requirements can be documented adjustments and statement adjustments in the quality manual. This adjustment is made by ISO 9001: 2015. From the results of the analysis, found ten clauses that are not by the requirements of ISO 9001: 2015 are clauses 4.1, 4.2, 4.3, 4.4, 5.1.1, 5.1.2, 6.1, 7.1.1, 7.1.6 and 9.3.2.

References

1. BSI. Moving from ISO 9001:2008 to ISO 9001:2015. London: BSI Group. (2014).
2. ISO. International Organization for Standardization 9001:2015 Quality Management. Geneva: ISO. (2015).
3. B. Purushothama., Effective Implementation of Quality Management Systems. India: Woodhead Publishing India Pvt. Ltd. (2010).
4. Psomas, E. L. (2013). The effectiveness of the ISO 9001 quality management system in service companies. Total Quality Management & Business Excellence, Vol. 24, No. 7, 769-781.
5. A. Lester., Project Management, Planning, and Control (6th Edition ed.). USA: Elsevier. (2014)
6. BSI. BSI Training Catalogue. London: BSI Group. (2017).
7. ISO. International Organization of Standard 31000:2009 Risk Management. Geneva: ISO. (2009).

AUTHOR INDEX

EDP Sciences – Web of Conferences
17, Avenue du Hoggar
Parc d'Activité de Courtabouf
BP 112
F-91944 Les Ulis Cedex A
France

ISBN 978-1-5108-7248-6

CURRAN ASSOCIATES INC.
proceedings.com

9 781510 872486

XI International Road Safety Seminar (GAMBIT 2016)

Road Innovations for Safety

MATEC Web of Conferences Volume 122 (2017)

Gdansk, Poland
17 - 18 November 2016

edp sciences